TREE STORIES

TREE
STORIES

How trees plant our world and connect our lives

Stefano Mancuso

Translated from the Italian by Gregory Conti

Profile Books

This paperback edition published in 2024

First published in Great Britain in 2023 by
Profile Books Ltd
29 Cloth Fair
London
EC1A 7JQ
www.profilebooks.com

First published in the United States of America as
Planting Our World in 2022 by Other Press LLC, New York

Originally published in Italian as *La pianta del mondo*
in 2020 by Editori Laterza, Rome.

Text designer: Jennifer Daddio /Bookmark Design & Media Inc.
This book was set in Vendetta Medium by
Alpha Design & Composition of Pittsfield, NH

1 3 5 7 9 10 8 6 4 2

Printed and bound in Great Britain by
CPI Group (UK) Ltd, Croydon CR0 4YY

A CIP catalogue record for this book is available from the British Library.

ISBN 978 1 80081 547 6
eISBN 978 1 80081 548 3

To Paola and Sonia

CONTENTS

Prologue   *1*

1   PLANTING LIBERTY   7

2   PLANTING CITIES   39

3   PLANTING THE UNDERGROUND   71

4   PLANTING MUSIC   89

5   PLANTING TIME   105

6   PLANTING KNOWLEDGE   131

7   PLANTING LAW AND ORDER   167

8   PLANTING THE MOON   187

Notes   *197*

PROLOGUE

After decades of keeping community with plants, I seem to perceive their presence not only in every place on our planet but also in the stories of each and every one of us.

At first, I figured that a heightened perception of the vegetable world was the normal consequence of my sensitivity to these silent beings. And that, as happens to anyone who develops a strong taste for something, I had started to notice the object of my interest everywhere I went. Anyone who has fallen in love knows what I am talking about: that strange sensation that everything in the universe, no matter how distant or marginal, appears to be related in some way to the object of our affection. Every event, every song, the weather conditions, the stones in the sidewalk you're walking on, everything has a precise echo in your own little love story. I remember a delightful novella by Guy de

Maupassant, which I read as a young boy, about a lady who, every time she fell in love, and she did so fairly frequently, radically transformed her world by placing at the center of her own interest the profession of her new lover. She fell in love with a lawyer and spoke only of civil codes and trials; with a pharmacist the world was composed solely of medicines and drugs; with a jockey it was all horses, saddles, bridles, and reins. I am sure that each of us knows of analogous cases. It is one of the reasons why somebody who's in love is unbearable.

So I started asking myself if it wasn't because of some kind of green infatuation that, like the lady in Maupassant's story, I couldn't see anything except plants all around me: in every place on the planet, at the start of every human story, the basis for every event. I have thought about it and I think I can assert with a certain degree of assurance that the answer is no. I am reasonably sure of it. That I live with plants, study them, and that they are undoubtedly the center of my interests is not related to their appearance at the start of every story. It is simply a consequence of their enormous number and of their being the source of life on this planet. This is an indisputable fact. How could it be otherwise? We animals are only 0.3 percent of our planet's biomass, while plants are 85 percent. It is obvious that every story that takes place on our planet has, in one way or another, a leading role for plants. This planet is a green world; it is the planet of plants. It is not possible to tell a story about it that does not stumble across its most numerous inhabitants. That plants do not show up in our experiences or, if they happen to creep into

2

them, they only have a role as colorful extras, is the fruit of our total removal from our perceptive horizon of these living beings on whom all life on earth depends.

When we succeed in looking at the world without seeing it simply as a playground for humans, we cannot help but notice the ubiquity of plants. They are everywhere and their adventures inevitably intertwine with our own.

The English composer Sir Edward Elgar was once asked where his music came from. His answer was: "There is music in the air, music all around us, the world is full of it and you simply take as much as you require."[1] The same thing applies to plants. They are, like music was for Elgar, literally all around us, and to write about them, all you have to do is listen to their stories and tell them, each time using *as much as we require*.

That is how this book came to be, by taking, here and there, stories of plants that, as they intertwine with human experience, bring both together in the narration of life on earth. Just like what happens in a forest, where each tree is linked to all the others in an underground network of roots that joins them together to form a superorganism, plants are the nervous system, the map (or plan) on which the world we live is built. Not seeing this plan, failing to use it in planning our world, believing that we humans have by now placed ourselves above nature, is one of the gravest dangers to the survival of our species.

Ficus macrophylla (Moreton Bay fig). This gigantic, spectacular, apparently invulnerable tree depends for its reproduction on a single insect, the *Pleistodontes froggatti* (fig wasp), without which the figs remain unfertilized and fall prematurely.

Like liberty, it is majestic and solid, but difficult to reproduce.

One

PLANTING LIBERTY

E ver since I can remember, I have had an irresistible
attraction to paper. When I was three, I fell in love with
my nursery-school teacher and right after that with paper.
This second passion has persisted unaltered and intact from my
childhood, accompanying me since well before I started taking
an interest in plants and such. One of my earliest memories of
emancipation is tied to paper. To be more precise, to paper in the
form of comic books.

At the time, I believed comic books came directly from
the generous hands of my parents or other adult relatives,
who at more or less regular intervals, and for reasons almost
always related to birthdays or the achievement of some hoped-
for result, bestowed on me, solely of their own volition, one
of those fantastic illustrated stories. Sure, I was aware of the
fact that comic books came from those places of wonder called

newsstands, those sacred structures to which only adults were admitted and so for me were as inaccessible as if they were located on Mount Olympus.

Then one day—I must have been seven—during a vacation in Rome, there appeared before me, totally out of the blue, the first secondhand comic-book stand of my life. Children the same age as me, with or without their parents, adult men and women, all allowed in to revel in the marvels of print, without discrimination of any kind. Not even income. The one hundred lire required for the purchase of a comic book (four hundred lire for five) was well within the range of my financial possibilities. In fact, I always carried with me a thousand-lire bill entrusted to me by my father "for any eventuality." I never had any idea what an "eventuality" might be before that day. I invested the thousand lire in twelve (consecutive) issues of *Comandante Mark*. It was a magic moment.

Since then, first for comic books and later for books, secondhand markets have been part of my daily life. I have followed some of them in Florence through changes of location and ownership and generations of operators, and although none of them has touched my heart like that first one in Rome, lots of other books discovered in secondhand markets around the world are indelibly impressed on my memory. Like the time at the Marché du Livre Ancien et d'Occasion George Brassens in Paris when I put my hands on a little book whose frontispiece bore the magnificent title *Essai historique et patriotique sur les arbres de la liberté*.

The market I'm talking about is one of those can't-be-missed occasions for all those who share with me the insane passion for flea markets and used books and who live, or happen to be spending a weekend, in Paris. Every Saturday and Sunday, fifty to sixty bouquinistes come together in and around the park and market named after George Brassens, in the fourteenth arrondissement, to display their merchandise to a robust number of fanatical bibliophiles. We recognize one another immediately. It's always the same crowd, we run into each other in the same places, all desirous, weekend after weekend, to begin searching the thousands of haphazard heaps plopped on the stands of the booksellers. There are those who for years have been on the lookout for the only issue they're still missing to complete their collection of some obscure series from the turn of the twentieth century; others collect books on the most improbable topics, like coffeemakers (I actually met one such collector), Finnish history, Japanese weapons, or microorganisms of the soil.

For the most part, the crowd is made up of academics who, having studied enigmatic subjects for years, end up getting trapped in the world of their research. Okay, I'm not all that different myself, I admit. I wander around those same stalls looking for any book of any kind about plants and trees, if possible published before the start of the nineteenth century. Here, in years of assiduous rummaging, practically every Saturday morning of my adult life when I've had the good fortune to be in Paris, I have gathered an imposing collection of abstruse, forgotten, and absolutely marginal books, whose only common feature is their subject matter: plants.

The market opens to the public every Saturday at nine. This means that the truly passionate collectors are already there at eight, lying in wait. They meet in a coffee bar right in front of the market, all outfitted with enormous empty knapsacks that they hope to fill. There are embarrassed exchanges of greetings between people who have known each other by sight for years, who often know each other's names and professions, but who have never really sat down for a chat. They drink their coffee and glance at each other suspiciously, especially if they are rivals with the same interests. It's a sort of curse, whatever the subject of your search, there is always someone who is your direct competitor.

Naturally, I have my antagonist, too. He is an elderly gentleman, as tall and thin as a reed. He has a wrinkly dark complexion as though he had been left to dry for years under the desert sun, always dressed in what looks to me to be the same long, light-colored trench coat, summer or winter. Insensitive to the climate, like all good book hunters, rain or shine, snow or gale-force winds, freezing cold or stifling heat, he is always there. Every Saturday at eight. He roams the stalls with a slight limp that he uses as a weapon to conceal his ferocity. You think he's got to be slow-moving, but instead, the instant something attracts his attention, he's capable of scaling enormous piles of books with the agility of a young boy. I know this by now, but among the neophytes, his apparent fragility slays countless victims.

Fragility? He doesn't know what it is. He's as sturdy as the seasoned wood he appears to be made of. He's got stamina, too, damn him! He never gets tired, methodically examines each and every pile, and nary a Saturday goes by that he doesn't go home with a knapsack stuffed with ponderous tomes. I once heard a bookseller call him *professeur* and another Henri. So Professor Henri was all I knew about my adversary, besides the fact that he was a terror, a real tough nut, who adores botany and the French Revolution. And that he irks me. He seems to have a sixth sense for botany books. He burrows into a pile of books like a weasel into a rabbit warren, and he always comes up with something in hand. Whenever we cross paths among the mounds, it seems like he's observing me with equal parts of disdain and amusement. We eye each other from a distance, and at the start of the day we normally head for opposite poles of the market, hunting for freshly unloaded piles, spying each other warily, hoping to be the first to find something of interest to the other. A tough life, I assure you.

It was during one of these close encounters with Professor Henri that I happened upon the famous little booklet. It was inside a plastic cover, like the ones we used when we were kids to protect our notebooks. I don't know if kids use them anymore, but when I was in elementary school, putting those covers on my notebooks at the start of the school year was something I liked a lot. For that and only that, in memory of childhood days and curious to find out what it was about, I picked up what I was convinced was no more than a notebook. I began leafing through it distractedly and, to my complete surprise, under that banal cover appeared a lovely leather binding from the late eighteenth century. Standing next to me, Professor Henri seemed to have, like a chameleon, two independent eyes—one reserved for his own search and one constantly fixed on me to monitor what I was up to. He noticed the eighteenth-century binding and froze as though he were paralyzed. I had him in my grip. With a deceit I didn't know I was capable of, I turned the book so that only I could see the pages and continued leafing through it, leaving him to be tormented by uncertainty. When I came to the frontispiece and could finally determine what the book was about, I took my revenge. Feeling like a poker player who had just been dealt a royal flush, I faked an expression of disappointment, to the professor's delight, made a feint of throwing the book back on the pile, and then, as though changing my mind at the last minute, called out to the bookseller a lackadaisical "I'll take it," paid what I owed, and set the book aside.

Professor Henri followed my every gesture. We went on leafing through books that popped up. Every now and again, I took another look at my little book and then, with an annoyed grimace, put it back down. In the end, he took the bait. He dipped into books all around it and then, unable to contain his curiosity, asked me politely, "Excuse me, monsieur, may I take a look at the book you have just acquired?" "Oh, certainly, please, as you like." The professor leafed through it quickly, came to the frontispiece, and stared petrified at the magnificent title: *Essai historique et patriotique sur les arbres de la liberté*, authored by a certain Grégoire. The professor couldn't take his eyes off the book; he leafed through it incredulously. I allowed myself to point out to him: "You see? It was published in the second year of the Republic, in 1794, if I'm not mistaken." I looked at him without managing not to smile. "Who knows how such an interesting book came to end up in a cover for notebooks." He was so taken aback that I regretted having been so cruel. To earn his forgiveness, and since it was the right time, I asked him if I could take him to lunch. I wanted to know something more about him. He accepted. So we headed off toward a nearby brasserie.

His full name was Henri Gerard and he was a professor of French history. I asked him to tell me some more about his passion. "I've been observing you search through books for some time now," I began, "but I thought your fancy was botany, not history. After all," I went on, "except for this last occasion in which I had more luck than you, it's been years that you've been humiliating me in the search for books on botany." That admission of his ability as a book hunter seemed to soothe, at least in part, his earlier mortification. His smile reappeared for an instant. "You're right about history and botany. They are really my two true passions." I pulled the book out of the bag at my feet. Now that I had removed the plastic cover, we could both see that it was in excellent condition. I opened it to the frontispiece. Only the author's surname was mentioned: *Par Grégoire, membre de la convention nationale* was the only indication. "Now," I replied, "you, who are an expert on the revolution, tell me something about this Grégoire, who writes about trees. Did you say he was an abbot?"

He looked at me askance. "You don't know who Henri Grégoire was, the *citizen* priest?" He was incredulous in the face

of such an enormity. "No, I've never heard of him," I replied
serenely. He took the little book from me and started turning
it in his hands, shaking his head. He couldn't bring himself to
believe that it had ended up in the hands of someone as ignorant
as me. "Well, by now it's yours and there's nothing we can do,"
he said with a sigh, carefully laying the book down on a corner
of the table. "You should at least know something about the
author of this work of which you have so undeservedly entered
into possession. I hope you will understand its value. Henri
Grégoire, better known as Abbé Grégoire, is absolutely one of the
most important and fascinating figures of the revolution, but I
had no idea that he had also written a book on the liberty trees.
What's more," he added, "from the publication date he must have
actually been the first to write about them." He gave me a long
stare, carefully sizing me up. "If this subject really interests you as
much as it interests me, why don't you come to see me at my home
this week? Choose whatever day you like, we'll have a coffee and
exchange information about the liberty trees. What do you say?"
I accepted the invitation immediately for the next Wednesday.
The professor thanked me for the lunch, gave me his address, and
without waiting even for the time it took me to write it down, left
me to head back to more piles of books.

I didn't follow him. I didn't feel like more rummaging through
books at the market, while I was very curious to know more
about the story of the liberty trees. So I decided to remain seated
comfortably at the table and devote the rest of my afternoon to
reading my Henri Grégoire.

Page after page he unveiled the mystery of the liberty trees:
First of all, that they were actually real trees. I was concerned
that the term was only a metaphor but no, they were real trees, in
wood and leaves. Trees that, during the years of the revolution,
had been planted in every inhabited place in France, from the
most minuscule villages to the capital, as a tangible symbol of
the revolutionary ideals. A magnificent custom, whose origin,
however, was rooted in an earlier revolution: the American
Revolution.

In 1765, the English approved the infamous Stamp Act,
which imposed a tax on every sheet of paper printed in the
American colonies. Printing paper had to be imported from Great
Britain and had to be stamped with a tax stamp certifying the
payment of the duty. In this manner, the British would be able to
control anything that came to be printed in the colonies and, at
the same time, the revenues from the tax would pay the expenses
of maintaining the army deployed in border defense. Protests
erupted immediately in all the territories controlled by the British
monarchy in North America. At first, cautious and peaceful,
and then more and more numerous and violent, the disorders
grew more intense to the point of becoming authentic forms of
rebellion against the Crown.

The most clamorous of these protests occurred in Boston on
August 14, 1765. A crowd of infuriated colonists gathered under
a giant elm tree and hung the effigy of Andrew Oliver, the Boston
merchant chosen by George III as the official in charge of the
implementation of the law. Also strung up was a boot whose sole

was painted green to represent the two ministers, Count Bute and Lord George Grenville,[1] thought to be the real proponents of the tax. This is the first public act of defiance of the English Crown, which would lead, ten years later, to the beginning of the American Revolution.

The elm under whose branches the Boston protesters gathered came to be known as the Liberty Tree, and the area around the tree where the protesters demonstrated was rebaptized Liberty Hall. When, in response to the protests, the Stamp Act was abrogated in 1766, the biggest celebrations took place under the elm tree in Boston, dressed up for the occasion with flags, ribbons, and lanterns. The tree thus became the best-known symbol of resistance to English rule and many other cities soon inaugurated their own local liberty trees.

But, as is often the case, being a symbol is not a passport to a long and peaceful life. In fact, during the siege of Boston in 1775 and 1776, upon the outbreak of the war against the English, the poor elm tree, which had been planted in 1646 and which, had it not become the Liberty Tree, would have easily survived for some centuries, ended its life precociously. It was cut down by the English and Boston's Tories and used for firewood. This Boston elm tree was the first of all the liberty trees.

But while the Americans were the first to adopt this symbol, its diffusion is without doubt to be attributed to the French Revolution. Abbé Grégoire writes in his book that the first person in France to adopt the tree as a symbol of liberty and fraternity was a certain Norbert Pressac, the pastor of Saint-Gaudens near

Civray, in the department of Vienne. In May 1790, Pressac "had a handsome oak tree extirpated from the forest and transported to the village square, where the two sexes helped to plant it." Then the pastor harangued the crowd with this phrase: "At the foot of this tree you will remember that you are French and in your old age you will remind your children of the memorable epoch in which you planted it."

The idea of planting liberty trees was embraced by the patriotic spirit of the French people and the practice spread rapidly throughout the nation. However, possessing your very own liberty tree is not always a simple proposition. Many villages do not have trees of dimensions large enough to fit the bill. Moreover, revolution or no revolution, since every town, road, and house wants its own personal liberty tree to *raise its majestic head* higher than the others, forests were besieged by search parties looking for the tallest specimens to be uprooted and turned into liberty trees. It must not have been an easy time for those we today call "vegetable patriarchs." For every large tree reasonably close to a population center, its fate was sealed. And since, as Grégoire again reminds us, "the desire to procure gigantic trees rapidly did not allow for the selection of well-rooted trees, their quick desiccation was the result."

To refurbish the scarcely dignified panorama of liberty trees dead for lack of roots, the Convention issues a decree by which "in all municipalities of the Republic in which the liberty tree has perished, a new one shall be planted between now and the first of Germinal. This planting and its maintenance shall be entrusted to the care of the good citizens so that in every town the tree of liberty shall flourish under the aegis of French liberty." The spirit of the decree is clear: A dead tree cannot be the symbol of an eternal revolution. "Dying or dead nature must be only the emblem of despotism," Grégoire reminds us. On the contrary, "living and productive nature, which reinforces and spreads its benefits, must be the image of liberty, which enlarges its dominion and brings to maturity the destinies of France."

The liberty trees must be majestic and perfectly healthy and possess a whole series of other characteristics. According to Abbé Grégoire, the perfect tree must:

1. be strong enough to withstand the coldest climate, otherwise a severe winter would be enough to eliminate it from the soil of the Republic

2. be selected from among the tallest trees, ranging from eighty to a hundred and thirty feet tall, because the strength and magnificence of a tree inspires a sense of respect that is naturally tied to the object of which it is the symbol

3. have a circumference that occupies an ample extension of terrain

4. have an extension of shade large enough to provide citizens shelter from the rain and heat under its hospitable branches

5. be long-lived and if it cannot be eternal, at least be chosen from plants whose life can go on for centuries

6. be able, finally, to grow in solitude in all the regions of the Republic

It is clear that not all trees satisfy all these requirements and that really very few of them possess the grandiosity necessary to represent worthily the greatness of the revolution. Abbé Grégoire has no doubt that only one species possesses all the necessary features. The liberty tree par excellence should be an oak.

Finished with my reading of the little book by Henri Grégoire, I finally had a fairly clear idea of what the liberty trees had represented for the revolution, but I did not feel at all ready to face the enormous erudition of the professor. I spent the days that separated me from our meeting trying to gather as much information as possible about this fascinating history. I read everything I was able to find on the subject, but without learning much more than what I had already picked up from the abbot. The impression I got was that not all that much was known about it and that everyone, to a lesser or greater extent, did little else but repeat the things I'd read in the essay by Abbé Grégoire, even if very few of them bothered to cite it.

So it was with an understandable dose of the anxiety, which I have carried with me since the days of my university exams, that I found myself on the appointed Wednesday near the home of the professor, well in advance of the agreed-upon hour. From his one and only trench coat, his tenacious bargaining over the sale price of books, and the story of his emigration and return to the homeland, which I had overheard at the book market, I had the idea that the professor was not well off. I imagined that he lived in a dignified but modest neighborhood. The street address where I now found myself, however, had nothing modest about it. On the contrary. Corresponding to the address indicated to me by the professor was an imposing apartment building protected by an even more imposing entrance, complete with telamons holding up the architrave. The grandiosity of the building was so at odds with the figure of Henri that I figured he had decided to get back at me by sending me to a fake address. The entrance was guarded by an enormous front door in wrought iron, and, no matter how hard I looked, I could not locate any doorbells, or an intercom, or any other system that might permit entry. The situation was becoming ridiculous. I had already tried a couple of times to knock timidly on the door. I was just about to leave when the door began to open onto a most refined atrium and an elegant doorman in full uniform welcomed me, asking if I was the guest expected by Professor Gerard. Trying to hide my surprise in the face of the unexpected luxury of the building, I replied that yes, I was. The doorman accompanied me to the elevator. "You'll find the apartment of Monsieur Gerard on the fourth floor. He is expecting you."

The professor was waiting for me in front of the entrance to his apartment, dressed impeccably in a tailor-made suit. The amazement on my face must have been so evident that he could not hold back an uproarious laugh. "Come in, my friend, and excuse me if you see me today in such different garb from what I was wearing the last time we met." I kept up my incredulous stare. "There is nothing that can't be explained, believe me. The version of myself as a book hunter that you met at the market is the real one. Don't let yourself be deceived by appearances. I can't in the least afford this apartment, and the suit I'm wearing is also the only elegant suit I own and I use it very rarely. But today felt like a happy occasion worthy of celebration: two colleagues with a shared interest in liberty trees. I'm sure that doesn't happen often. This apartment was left to me as an inheritance by my parents. My mother's family once owned the entire building. Then this apartment was all that was left . . . and I'm lucky I don't have to pay the condominium fees. I could never afford them on my high-school teacher's pension. Well," he continued, smiling at me, "I hope this gets you talking again . . . otherwise, I'll have to hurry up and change my clothes."

I shook myself out of my stupor and muttered some words of apology. Meanwhile, Henri led the way into a monumental apartment that, on first glance, contained nothing but books. We walked through an endless string of empty rooms. Aside from their floor-to-ceiling bookshelves loaded with books, there was not a single chair, table, or painting, or any of the usual household furnishings. Only books. Thousands and thousands of them. As we were walking, the professor rapidly indicated the subjects in which his vast library was subdivided. "In this room and the next, ancient history. Then the Middle Ages, Renaissance, the Enlightenment. Here, travel, there, geology," and so on, room after room, subject after subject, until we came to a large sitting room. Finally, a long table, several chairs, and two enormous armchairs. On one side of the room there was a fireplace and, above the fireplace, in the only space in the house not occupied by books, a large painting of a gentleman in ecclesiastical garb sitting at a desk, whom I figured must have been an ancestor.

The professor gestured me to one of the armchairs and invited me to sit down. He hadn't yet sat down in the other one before he was asking me impatiently, "So? Don't keep me on pins and needles. What have you discovered about our beloved liberty trees?"

I started telling him about the contents of the book. I had prepared myself for the exam and my summary was replete with data and references. All the same, nothing I was saying seemed to be of the slightest interest to him. He kept nodding as though all my information were old news. The only novelty was the booklet itself. He hadn't been able to find any reference to this work by Grégoire. None of the authoritative sources he had consulted, including the catalogue of the National Library of France, listed the abbot's essay. This, however, was not as rare an event as one might think, especially for works published in the early years of the revolution. In that period, short essays of a few pages, like the one that had fallen into my hands, were rather common. They were used as manuals to be consulted by the citizenry. "These booklets," the professor began, "contained practical directions, often to be followed with complete fidelity to the text, for the achievement of objectives to which the Convention had given top priority." What the professor found least convincing of all the indications reported by Abbé Grégoire was the requirement, repeated several times in the text, that liberty trees had to be enormous.

"I don't understand why they had to select such big trees," the professor continued, shaking his head. "Majestic and long-lived, fine, but obtaining really large specimens must have taken a huge effort and substantial resources." He couldn't be persuaded, couldn't believe that the Convention had required such an endeavor for a simple symbol. "At the time, the level of difficulty must have been colossal," he went on. "Imagine having to uproot a tree eighty to a hundred feet tall, presumably weighing up to several tons. They would have had to dig enormous holes in order to preserve the precious root network. And how did they manage to keep the uprooted trees from falling to the ground? How many people did they need to bring off an operation of that magnitude?" As he spoke, he kept shaking his head, with increasing frequency, to underline his incredulity that all of this was possible in the years of the revolution. "And that's not all," he continued. "The trees had to be transported long distances, even several kilometers, through forests without roads. I don't see how such a thing was possible."

The difficulties were indeed enormous. "Anyway, Professor, we have to be careful not to make the mistake of believing that people in the past were not capable of performing huge tasks. Just look at the pyramids. Even without the means we have today, people have always been capable of accomplishing great things."

Henri didn't seem convinced. "I don't know. You're right about the past, but in this case there's something that doesn't ring true. Can you imagine the Convention forcing thousands of people to take on such huge tasks for a symbol? It was a time when the people were fighting for their own survival and for the ideals of the revolution."

He paused and went to one of the bookshelves to get the ordinances of the Convention in 1792. "Look here: food, transportation, hygiene, militias, animals, canals, defense, borders. These were the things they had to take care of. Certainly not the moving of gigantic trees from the forest to the city centers. The more I think about it, the crazier it all seems. And keep in mind that we're talking about an ordinance of the Convention! It would not have been tolerated that the revolution was represented by dead trees planted in the village center. Really, I don't understand it. Today, when we have immensely powerful machines, the survival rate of transplanted large trees is extremely low. Only a few specialized companies are able to guarantee good survival rates. At that time, there must have been innumerable trees that died and those responsible would have been hard-pressed to explain why."

He was right, the same thought had occurred to me. "A few years ago," I explained, "a good friend of mine showed me some images documenting the operations necessary for the transplanting of large trees from a forest to the gardens of a villa owned by a Russian magnate. Even though the trees had to be transported for an average of less than five kilometers, new roads had to be built in the forest that were wide enough and robust enough to allow for the passage of the trucks and machines involved in the transplanting. They had to use scaffolding, cranes, steam shovels, and trucks dozens of yards long to transport the trees from the forest to the villa. At the time of Abbé Grégoire how could they have moved thousands of large trees from the forest to the centers of populated towns and villages and replant them in a way that ensured their survival? It must have been an extremely complex operation requiring the employment of huge numbers of people. So how do we explain that not a trace of these enormous undertakings is to be found?" The professor intervenes: "You're right! I hadn't thought of that. For the transport of those enormous trees throughout France, you would expect to find all kinds of prints produced to commemorate the enterprise. But I can't recall ever seeing even one. There are hundreds of drawings regarding the liberty trees, but none that show their transportation from the forest to the towns." He got up again and without hesitation pulled down from a shelf a large volume of prints from the revolutionary period. "Take a look at these while I make some coffee. Maybe we'll get lucky." He handed me the volume and went off in the opposite direction from where we had come.

I started leafing through the book distractedly, slowly turning the pages one by one and devoting only slight attention to the succession of prints. I was much more interested in looking around the room. How big was that apartment? By my calculations, just the room where we were sitting had to be much bigger than my entire apartment in Italy. At Paris prices, owning an apartment that size and in that neighborhood made the professor a very wealthy man ... yet there was something strange about it all. The total lack of furniture, the less than satisfactory level of cleanliness, his clothing, except for the luxurious suit he was wearing today, all told a different story. Much more modest. And those books. How could it have been possible to buy them on a high-school teacher's pension?

These were the thoughts chasing around in my head when, on one of the pages of the ponderous tome I was leafing through, I saw something that caught my attention. It was a map of Europe and the Americas and, if it hadn't been for the caption at the top, "Fraternity Trees," and the date, 1848, I probably wouldn't have tarried over it another second. At first glance, it didn't have any noteworthy characteristics, yet its title was clear: In some way that map concerned the subject that was dear to our hearts. Despite examining it attentively, however, I didn't see anything of interest. Everything was in its place and nothing seemed any different from an ordinary geographical map. But then, why that title, "Fraternity Trees"?

I brought the map up closer to my face to get a better look. It seemed like I could glimpse something like a thick network of lines uniting the villages and cities on one side and the other of the Atlantic. I felt a chill run down my spine. I reexamined the map with more calm; I had to be sure. It really resembled a root network, but the map I was looking at was only a reproduction and not even very detailed. Maybe they were only wrinkles in the paper, shadows, or some lines with a completely different meaning. Maybe I was so used to studying root networks that I was beginning to see them everywhere. Yet the more I examined that map, the more it seemed to me that the roots extended to connect the place-names. I decided to think of something else; I would ask the professor to tell me what he saw. I started getting the sensation that sometimes comes over you when you're watching clouds go by and suddenly an image emerges that is so clear and precise you can't imagine that other people can't see it, too, or that they see something else.

The professor called to me from the kitchen. "Come and give me a hand, please. I can't carry everything out there by myself." I helped him bring the coffee and everything necessary to serve it and we sat back down in our original places.

"Did you find anything?"

"Something, maybe, but I'd like to hear what you think of it." I opened the book to the page with the map and handed it to the professor.

"'Fraternity Trees,'" he read at the top. "They're the same thing. Liberty trees or fraternity trees—they are referred to with both names indifferently." I waited anxiously; I didn't want to be the one to suggest what to look at. "But I see absolutely nothing of interest. The thing I don't understand is why this map has this title. You'd expect to see a map with the location of the trees and instead there's nothing. A perfectly ordinary map." He looked at me quizzically. "What was I supposed to see?"

"Take a closer look, please." I didn't want to influence him in any way.

The professor put his glasses on and moved closer to the map to check the details. Suddenly, a look of surprise came over his face. He started shifting his gaze from one side of the map to the other, moving in closer and then farther away. More and more puzzled as the more careful examination revealed something that he hadn't seen before. "You mean these thin lines, right?" I nodded in satisfaction; he could see them, too. "They would seem to be a sort of road network that joins the towns and cities of the two continents. But obviously they can't be roads. Look here, for example, dozens of these lines cross the Alps in places where I don't think any passes or trails exist. And then in some areas the lines are much more dense than in others, even though they are regions where the arteries of communication must have been very scarce in 1848. Look here, in southern Italy"—I moved my chair closer so we could study the map together—"there is a thick network of lines running right through Calabria. But Calabria had very few roads at that time. Anyone who wanted to go from Naples to Sicily went by ship, precisely because it was so difficult to find practicable roads."

The professor got up and went into another room to get a road atlas from 1890. The resources of that library were infinite. "This I got at the George Brassens market," he said with a smile. He leafed through it until he came to a map of southern Italy. "Here we are. This shows that in 1890 Calabria had very few roads . . . meanwhile our network in the same area is very thick, even though it refers to fifty years earlier." He raised his eyes from the map and looked at me confused. "What do you think these lines mean?"

While he was speaking, a crazy idea went through my head, but I didn't want to talk about it yet. I needed to think about it some more. So I kept my response vague. "To me they look in every way like roots joining regions, cities, and towns on one side and the other of the Atlantic."

"Roots?" He looked at me like I was out of my head.

"Yes, roots, and don't look at me like I'm mad. I've spent my whole adult life studying roots. Believe me, I know a root network when I see one. And the one here on this map is undoubtedly the topographical representation of a root network. Or, if you really want to know, I believe it is the representation of a network of root networks."

"A . . . what?" the professor almost shouted.

"It has the exact shape of the representation on paper of an underground network of the roots joining the trees of a forest."

Now the look on his face was more attentive. "Could you explain that better? Keep it simple, please, despite my passion for plants and botany books, my knowledge of how plants work is very rudimentary."

"Well, this should rightfully be a rudiment. The trees that are part of a forest or a wood are not separated from one another. Rather, by way of their roots, they form an underground network that joins them all together in an enormous widespread network.

A forest, in other words, should be seen as one superorganism born of the interaction among the trees that are part of it. A little like what happens in an ant colony. There are countless ants in a colony but the entire colony behaves as though it were a single individual. The same thing happens with trees." I looked at him. "Does that not suggest something to you?"

The professor again looked quizzical. "What should it suggest to me?"

"The fact that trees are joined together in a community. Or in a 'confraternity,' to use a term closer to your customary field of study."

"So, if I understand what you're telling me, these maps would concern a confraternity of trees—"

I interrupted him. "Not simple trees but liberty trees. In fact, I don't think it is a coincidence that the map uses the less common name 'fraternity trees.' Maybe I'm wrong, but it seems to me that the maker of this map actually wanted to point out to us the greater value and benefits of a confraternity of individuals compared to the same number of single specimens. When you think about it, the motto of the French Revolution, *Liberté, égalité, fraternité*, though never fully realized in any real human community, has, on the contrary, been perfectly realized in vegetable communities that live in a regime of perfect sharing through these networks that unite them—"

The professor interrupted me immediately; he was much less disciplined than I in the guise of the student. "But if what you're saying is true, it should be very simple to verify."

"How?" I asked him.

"To start with, if this map actually represents the relationships between the fraternity trees, it seems obvious to me that every node of the network marks the place where a tree was effectively planted during the revolution."

"Not only during the Revolution of 1789," I intervened. "Look at the date: 1848. I don't think it's a coincidence. I've read that the 'revolutionary' practice of planting liberty trees or fraternity trees continued with even greater enthusiasm during the uprisings of 1848."

While I went on talking the professor stopped paying attention to me and started looking for some book among the endless rows of bookshelves. He moved around the room from one side to the other, following sections and leads that only he was able to understand, murmuring the names of authors and book titles that passed before his eyes. He pulled out one book after another, leafed through them quickly, and then nervously put them back in place. "I remember very well having seen someplace a list of the liberty trees planted up to 1792. But I can't manage to find it." He kept roaming around the room until, with a liberating "Finally," he pulled down a book from the early twentieth century with a list of towns and cities that had certainly planted liberty trees.

By now, the number of books we were consulting was becoming unmanageable while sitting in armchairs. The professor suggested we move to the long table to work more comfortably. "So, I'll give you the place-name and you look to see if it's on the map and, above all, if it is a node in one of the root networks." He started reading off a series of village names. I immediately suggested that we switch roles. My knowledge of French geography wasn't sufficiently detailed for me to know the location of the minuscule hamlets that the professor had started to list. "I don't know where they are either. Better to use the Internet to find their correct locations." The professor read off his toponyms and, after pinpointing their exact position on the Web, I tried to locate them on our map. It was soon evident that this system didn't work. In its original form, the map was more than five feet long and three feet wide. It had been greatly reduced to make it fit onto a double page of the volume and a lot of place-names were illegible. In order to work effectively, we had to find the original or a far more detailed reproduction. Otherwise, we would never be able to see the correct names of the localities and ascertain whether the lines that we could just make out on our copy really represented a root system. In order to achieve any certainty that this whole story was not just the fruit of my imagination, we had to be able to study the original. Although the volume itself had no information that could help us work our way back to the original map, the professor was confident that we would be able to find it quickly. "You'll see, the National Library will be able to tell us something or, better yet, the Carnavalet Museum, which has an excellent collection of prints from the revolution. An old friend of mine works there, I'll call him in the morning and let you know."

The next morning—I hadn't even finished breakfast—I got the call from the professor. "Please, meet me at the Carnavalet as soon as you can. I've found the original. It appears that the mapmaker really did want to draw an enormous global root system. Hurry up, I'm anxious to hear what you think." I rushed, and in less than half an hour, we were sitting before a print of the fraternity trees. Its dimensions were much larger than the minuscule reproduction that we had tried to study first. Now, what had looked to us like almost imperceptibly thin lines joining different regions and cities, turned out to be roots without any doubt. Besides, the reproduction that we had consulted at the professor's house was missing the entire upper piece. In the original, which we now had before us, the image appeared in its entirety: It was an enormous poplar tree whose base was the starting point for the root system that connected the continents of Europe and the Americas.

"It is, for all intents and purposes, a map, or if you prefer a plan," the professor commented, content with his play on words, "that represents all the liberty trees planted up to 1848."

"No doubt about it. And also a very detailed plan, I would say. Now that it's easy to follow the roots on the map, you can see that the places connected by the system number in the thousands."

We got right down to work and, using the list that the professor had brought with him, went on with the operation we had started the evening before. Henri read the list of place-names where a liberty tree had been planted, and I checked whether the roots on the map created a node in correspondence to the place-name. In short, it became evident that we were looking at a map of all the world's liberty trees. Not only were their places represented on the map, but the thickness of the roots connecting the various localities was greater in proportion to the number of fraternity trees planted in a specific region. A fascinating representation of the functioning of a network, like so many that can be seen today in the research of anyone interested in complex systems,[2] but in 1848 it was decidedly cutting edge.

"Did you recognize the tree that the root network we are studying belongs to? It's a poplar."

"And so?"

"I don't think this species was chosen by chance. The Latin name for poplar is *populous*, people. The maker of this print wished to emphasize the tree's symbolic value. In sum, the plan of and for the world is the plant of the peoples that embraced the spirit of the revolution."

That day we finished checking the list of the liberty trees and, thanks to the map, were able to add a host of others. According to the list we put together, hundreds and perhaps thousands of liberty trees were planted in Paris alone. They were distributed in such a way that there was practically always one nearby. Every square, open space, parklet, or sizable courtyard had one. Even the suburbs of Paris had to have been covered with liberty trees. In France, as of 1792, an enormous number of trees had been planted. Abbé Grégoire writes: "In every town you see magnificent trees that raise their heads in defiance of the tyrants; the number of these trees comes to more than sixty thousand, since even the smallest villages are adorned with them and many of the large towns in the departments of the Midi have them on every street or even in front of the houses."

Of these fraternity trees that for a period of time united the places of the revolution in an invisible network only a few have survived, all hidden in out-of-the-way corners of Europe. In the great cities like Paris, for example, they no longer exist. Because they were such visible symbols, they soon became the targets of reprisals. They were mutilated, cut, lacerated, engraved with royalist inscriptions. The Convention having made them the object of continual legislative provisions made them one of the obvious emblems of a regime hated by many. As early as 1800 (Year VIII of the Revolution), only a few were left. During the Consulate or the Empire, those few survivors were rebaptized "Napoleon Trees," only to be definitively eliminated during the Restoration.

Some liberty trees were replanted in 1848 and still others during the brief experience of the Paris Commune in 1871. But each time the political regime changed, it was the liberty trees that paid the price: They are easy to cut down, they make a big noise when they fall, and they don't put up fierce resistance. So the survivors of the season when the trees united the peoples are really very few, unregistered, and generally located in villages or backwater towns in France or Italy. In Calabria, for example, there are some that managed to survive the Bourbon Restoration, or even worse, unplanned urbanization. But they are disappearing there, too. Soon there won't be any left. It would be useful to protect them and to tell their stories before the last liberty tree that we're able to see will be the one engraved on the French two-euro coin.

Unlike the ideal cities represented in Renaissance paintings, filled with architectural constructions and without the slightest trace of a single leaf of grass, the cities of the future will have to be completely covered with plants.

Two

PLANTING CITIES

The story of the liberty trees that once punctuated the panorama of many of our cities, large and small, comes to mind every time I happen upon one of the three magnificent Renaissance paintings known as *The Ideal City* in which, on the contrary, of trees there is not so much as a shadow. The paintings are conserved in the National Gallery of the Marche in Urbino, in the Walters Art Museum in Baltimore, and in the Gemäldegalerie in Berlin. These three famous paintings, whose artist/s is/are unknown but which are certainly Italian, represent the ideal of the perfect city. If you examine them carefully, you will notice that none of the three paintings has a trace of vegetation, not counting the few meager plants used as decorative elements in the Urbino painting.

Let's take this Urbino painting as an example. Attributed by many to Leon Battista Alberti—father of Renaissance architecture and author of *De re aedificatoria* (*On the Art of Building*), the fundamental architectural treatise of humanistic culture—the painting shows a piazza in central perspective with a magnificent church in the middle. The piazza is quite large and has a geometric pavement that transforms it into an enormous chessboard on which the buildings, like game pieces, are placed at regular intervals. The round church a perfect and concluded shape, the two symmetrical octagonal wells, the relationships between the dimensions of the buildings—everything in this city appears to be a pure representation of human thought. One might object that since these are paintings and not actual cities, they are merely representations. True. But even though never actually built, they are, nonetheless, the manifesto of how we imagine a city should be.

So, let's ask ourselves: Which has more influence on the structure of our cities, what we think a city should be or what purpose the city is supposed to serve? Although the inevitable response is that both aspects are important, I believe that cultural heritage, including, in a certain sense, our adaptive memory regarding how our home must be constituted, plays a prevailing role. Much of our ancestral memory concerns the necessity to defend ourselves. From the moment that the first human felt the urge to build a hut in order to settle in someplace, the inevitable consequence of this decision was to trace out a separation between his refuge and the nature surrounding it. Defense against predators, whether animals or humans, has always been an essential element in the construction of our settlements. The separation between outside the city, where nature reigns supreme, and inside, where, on the contrary, nature is completely excluded, is an ancestral remnant of those distant times.

The ancient city had to have walls and other mechanisms of defense that kept the inside of the city separate from and defended against a threatening outside. The presence of this impassable perimeter, in turn, meant that the dimensions of the urban interior couldn't be very great and that the extensive productive enterprises, farming, for example, with no space allocated to them inside the city walls, were bound to be situated outside of the inhabited center. What cities of every kind and every time have in common, according to Arnold Toynbee, an English historian from the early twentieth century, is that the inhabitants of a city are not able to produce, within the confines of the city itself, the food they need to survive.[1] A city is, therefore, necessarily separated from the natural context that hosts it. It is something very different from nature: It is the place of humans. A place created by us and in which nature is not allowed.

But is the form of the city as we have known it the only one possible? Can't we imagine in a different manner that which by now must be considered the home of our species? Up to now, we have left this imaginative exercise exclusively to the architects, while I believe it is essential that it become a mental arena for all of us. Indeed, how we imagine our cities in the near future will determine a substantial part of our chances for survival. Just to take one example, the form, materials, and functioning of our cities will have an impact on our chances of overcoming the challenge of global warming.

To understand this relationship, however, we have to approach this question from a wide perspective.

Today, humans are less and less global inhabitants of this planet. They had been such over the course of their history, even quite recently, when human populations could be found in every remote corner of the earth. But not in our time. Today humans occupy only a minuscule part of the surface of the planet, the part occupied by cities. In 2050, 70 percent of the human population—which will be close to ten billion people—will live gathered in urban centers, many of which will host tens of millions of inhabitants.

The astonishing rapidity of this phenomenon makes it very hard to fathom. In 1950 more than two-thirds (70 percent) of the people in the world still lived in rural settlements. In 2007, for the first time in history, the global urban population was greater than the global rural population and, since then, the velocity of the change has only increased. In 2030, the forecasts say that 60 percent of the world's population will live in urban areas and, in 2050, the percentage will rise to 70 percent, reversing in the span of a single century (from 1950 to 2050), the rural-urban distribution of the global population.[2] Obviously, with important differences in the different parts of the world. At one extreme is

Africa, which will continue to have a diffuse and rural population. At the other extreme are the Americas, in which more than 80 percent of the population live, already today, in cities. In Italy, the portion of inhabitants in urban areas comes to 71 percent, in Germany it is about 75 percent, while in France, Spain, and Great Britain it is well over 80 percent.

What is most striking about this high-speed acceleration toward urbanization is that it is in total contradiction with the rest of our activities. Communication, commerce, nutrition, industry, culture, and whatever other forms of human expression that might come to mind have tended to assume in modern times a universal and diffuse character, while, on the contrary, the choice of a place to live has been reduced, more and more, to a negligible portion of the earth's surface. Excluding Antarctica from the calculation, cities taken altogether cover an area that amounts to 2.7 percent of the dry land on the planet.[3] Their irresistible attraction is leading, on the one hand, to the depopulation of enormous areas once inhabited by humans and, on the other hand, to the concentration of the human population in densely populated places.

The point I find most interesting about this is that humans, in just a handful of years, have been reversing their own atavistic species behavior. The conquest of new territories has been the major occupation of our species since its appearance. For hundreds of thousands of years, we went in search of new territories to inhabit, pushing out from Africa to every other place on the planet. Then, in just a few decades, everything suddenly came to a halt. Let's take, for example, the story of space exploration. In 1969, we set foot on the moon for the first time... and, practically speaking, we've never gone back. Commander Eugene Cernan, pilots Harrison Schmitt and Ronald Evans, along with five mice, are not only the last men (and mice) to have gone to the moon since December 1972 but the last living beings to have gone beyond low earth orbit.[4] The conquest of the moon seems to have marked the apogee of human expansion. For the first time, a new territory has not become part of our habitat. For the first time, we have not had pioneers. For the first time in the history of human exploration, we have not returned, for more than fifty years, to a place we have explored. Our expansive push has been exhausted. No one seems to be interested anymore in colonizing new territories, while everyone feels an invincible attraction to amassing themselves in urban centers.

What does this behavior depend on? Alternating phases of expansion and contraction are normal in the geographical diffusion of all living species, whether vegetable or animal. Perhaps humans are going through a phase of contraction. We are used to thinking of ourselves as outside of nature, but we respond to the same fundamental factors that control the expansion of species: climate, changes in the ecosystem, interaction among species, abiotic factors, etc. It is very simple: The more favorable the conditions, the greater the diffusion of the species and, therefore, the greater its chances of survival. This affirmation shouldn't be surprising. Let's imagine that a species, formerly distributed throughout the entire planet, limits its presence for some reason, known or unknown, to small, delimited zones of the terrestrial surface. It is clear that the risks to the survival of this species will increase.[5] It is much more probable that some change incompatible with the species' survival should occur on the local level rather than the global level.

Organisms capable of colonizing environments that are very different in terms of climate, availability of nutrition, presence of predators, etc., are known as "generalists," while the others, those in need of specific environments in order to survive, are called "specialists."[6] Obviously, the chances of survival for generalist species are much higher. When environmental conditions change, the generalists manage to adapt better than the specialists that, instead, tend more easily to extinction.[7]

Just to take one example, think of the different capacity for survival of an omnivore, which can nourish itself on a variety of animal or vegetable foods, compared to a monophagous animal like the koala, whose only food are the leaves of eucalyptus trees. Note well, however: It is not only diet that dictates whether a species is generalist or specialist. A cactus, able to survive at high temperatures and in conditions of scarce water, is an example of a specialist vegetable. Within its restricted environment, in fact, it is very competitive with other species, but on the outside, it is not able to survive.

Judging from the parabola of our geographic expansion, it would seem that humans, once a generalist species, able to colonize any environment, are transforming very rapidly into a specialist organism, capable of thriving only within particular habitats with the characteristics of our cities. Indeed, regardless of their histories and geographical position, cities always have common characteristics that are absent in rural areas. With just a bit of an overstatement, we could describe our cities by applying the ecological concept of the *niche*.

In its original formulation, elaborated and disseminated by Joseph Grinnell, an American biologist and zoologist at the start of the twentieth century, the ecological niche of a species is determined by the fundamental characteristics of its habitat and by the adaptations that the species makes in this habitat to multiply and extend itself.[8] For example, going back to our cactus, its niche is defined by the combination of its desert habitat and the plant's physical traits and behaviors that allow it a positive adaptation. Now if we apply this definition of ecological niche to our cities, we realize that urban centers are becoming for us exactly what the desert is for the cactus: the only place where we can see ourselves prospering and multiplying, since it is the only place within which our specialization allows us to have the best chances of survival.

So in the urban environment the efficacy of our actions, calculated in terms of productivity or income (a highly debatable criterion, I realize), seems to be much better than in any rural environment. Evidence of this is provided by the relationship between the increase in per capita gross domestic product (the famous GDP) and urbanization. In 2008, in a survey involving 180 countries, a 10 percent increase in urbanization was associated with an increase of 61 percent in per capita GDP.[9] Within the urban niche, not only productivity but also the efficiency and availability of almost any activity one examines improves considerably. In cities almost everywhere, health and hygiene services, access to drinking water, the availability of transportation, schools, hospitals, etc. are superior in comparison with rural environments.[10]

The advantages of this specialization have not been slow in coming. If we look at any chart showing the growth in human population over time, we notice that a significant demographic increase begins about two centuries ago. In 1800, after thousands of years in which human population growth had been so slow as to be practically imperceptible, suddenly something changes, and the population mushrooms to a billion people. Since then, the acceleration has been continuous. To reach 2 billion it took us 127 years, 33 years to reach 3 billion, 14 years to go beyond 4 billion, and so on. An irresistible growth, one of whose causes is the industrial revolution, but it's not the only one. Urbanization and the development of the modern city have promoted demographic growth as much as the industrial revolution.[11] Indeed, the two factors are interrelated. They are two manifestations of the human specialization in the urban environment.

So all is well, then? In part. While the advantages of city living are numerous and unquestionable, it is equally true that the concentration of the largest part of the human species in such peculiar environments leads to risks that must not be underestimated. The reason is the same one for which the koala is much more susceptible than mice are to extinction: extreme specialization. While, on the one hand, the transformation of our species from generalist to specialist is advantageous in terms of access to resources, efficiency, defense, and distribution of the species, on the other hand, it exposes us to terrible risk. In fact, if those same urban conditions that allow us to prosper were to change, this would have a significant impact on our chances for survival.

Actually, it is a very simple matter. The specialization of a species is efficacious only in a stable environment. In changing environmental conditions, specialization becomes dangerous. Our urban success requires a continuous and exponentially increasing flow of resources and energy. Without a continuous intake of oil, gas, drinking water, wood, aluminum, iron, copper, lithium, tungsten, phosphorous, potassium, nitrogen, cobalt, ruthenium, molybdenum, lanthanum, lutetium, scandium, yttrium, neodymium, etc., the growth and maintenance of our species could not be assured. Now, many of the resources I have listed above, and very many of those I haven't, are (rapidly) depleting. I imagine that when they really start running out we will find replacements or will start to do everything necessary to recycle them. In a certain sense, although pressing, this is not the most important point.

What will change the environment of our cities definitively is global warming. A phenomenon so sudden and of such proportions as to represent exactly that survival-threatening change of conditions that we have been talking about. And, paradoxically, this change has its chief source right in our cities. Cities are, in fact, the main engines of our aggression against the environment. Currently, about 70 percent of global energy consumption and more than 75 percent of worldwide consumption of natural resources are chargeable to cities, which produce 75 percent of carbon emissions and 70 percent of waste. By 2050, cities will be called upon to house another 2.5 billion people, with a consumption of resources which, at the moment, is hard to imagine. In the face of these numbers, it is obvious that any solution to the problem of the human impact on the environment will have to pass through our cities.

But what could these solutions be? Fortunately, there are a lot of them and they will change every aspect of the way our cities function: from transportation to water consumption, from the production of waste to carbon dioxide emissions. Everything will be brought within closed cycles that will make the functioning of cities much more efficient. The solutions exist and, however slowly, they will succeed in limiting the damage.

What we can no longer put off is a deep rethinking of our idea of the city. The place where humanity lives—our only home—cannot be imagined as something separated from the rest of nature. To put it bluntly, those science-fiction images of an earth completely covered by enormous cities of steel and glass, without a trace of green and with streams of flying cars, are completely misguided and represent a dystopian and impossible future. Cities like the Los Angeles of *Blade Runner*; or Trantor, the planet-capital of Isaac Asimov's first galactic empire; or, to remain on the level of high culture, the Ecumenopolis which, according to the Greek architect and city planner Constantinos Apostolou Doxiadis, would fuse together, in a distant future, all of the world's megalopolises—these have no chance of ever seeing the light of day. They are obviously unsustainable. In order to survive, cities need constant flows of energy and materials, that is, resources that must be produced somewhere else.

To clarify the question of the resources needed by a city, we can use the concept of *ecological footprint*. The concept's inventors, Mathis Wackernagel and William E. Rees, describe it as "an accounting tool that enables us to estimate the resource consumption and assimilation requirements of a defined human population or economy in terms of a corresponding productive land area."[12]

Put differently, the ecological footprint measures all the resources (fuel, electricity, water, etc.) used by the people in cities and all their waste products, and converts it into a quantity of land (measured in hectares) needed to create those resources and to eliminate those waste products. The ecological footprint can refer, therefore, to a single person, a business, or to an entire city or country. Let's try to make a simple calculation. Multiplying the inhabitants of London—about 9 million—by the 5.4 hectare ecological footprint of a single Londoner, we obtain the result that the ecological footprint of the city of London, that is, the land area it requires for its needs, comes to 486,000 square kilometers (187,645 square miles).[13] This is about double the entire land area of Great Britain.

In the final analysis, cities, whatever their dimensions, can develop only because in some other part of the planet there exist natural resources that are exploited to feed their development. It becomes obvious that there must be a limit to the dimensions of cities. Sooner or later, the expansion of cities will stop. And the reason this will happen is the same reason for which limitless economic growth is unimaginable: because a finite planet does not have infinite resources. As much faith as we might have in what Leopardi ironically termed "the magnificent and progressive destiny" of humanity, the needs of urban populations will require ever more enormous amounts of resources that will have to be obtained from somewhere on our limited planet.

In other words, the expansion of cities happens at the expense of the natural resources of the planet. Let's take as an example the case of agriculture. For nearly all of the 300,000 years of the history of *Homo sapiens*, the earth was a place covered with plants. Forests or savannas occupied almost the total amount of habitable lands. As recently as a thousand years ago, it is estimated that less than 4 percent of the nondesert land above sea level and free of ice was used for the agricultural production of food.[14] Today, excluding the 10 percent of land area covered with ice and the 19 percent of sterile land (deserts, saline marshes, beaches, rocks, etc.), 50 percent of the remaining land is used in agriculture. The temperate forests, which in the eighteenth century covered more than 400 million hectares,[15] have totally disappeared[16] and even the tropical forests are in rapid decline. On the planetary level, forests cover a meager 37 percent of habitable land. Practically speaking, in just a few centuries, enormous expanses of forest have disappeared to make room for agriculture.

We have to ask ourselves if, in order to produce the food we need, we really need 50 percent of the usable land on the planet. The answer is no. Some 77 percent of this enormous area that we have subtracted from forests and natural ecosystems is used today to raise cattle and only 23 percent is used for the production of vegetable foods.[17] If we consider that raising livestock produces only 18 percent of the calories and 37 percent of the total protein consumed by humanity, this approach to land management seems nonsensical and totally illogical. We do not think enough about the enormous quantities of resources needed by cities at the current rate of urbanization. That most of the human population lives in cities is a modification of our living conditions that can only be compared to the change that happened about 12,000 years ago during humanity's transition from the nomadic life of hunter-gatherers to the stationary life made possible by agriculture.

I believe it is precisely the extreme rapidity with which this transformation has come about that has prevented us from understanding exactly what a city is: a highly structured environment, whose study has been hampered by an oversimplification of its complexity. For centuries, generations of city planners have believed they could govern and direct its development on the basis of a few simple design assumptions—whose efficacy could be, at best, extremely marginal—which, in any case, envisioned humanity as the only interested party. As was written as recently as a few years ago by Robert Beauregard, emeritus professor of urban planning at Columbia University: "In the practice realm of planning theorists, humans are the only meaningful actors ... Nonhuman things are not allowed the same ontological status as humans. Rather, they are presented as passive material objects to be manipulated through regulations, informal agreements, and incentives. In communicative action, legislation, and in theories of the right to the city, nonhuman things are epiphenomena. Only human beings have theoretical significance."[8] There can be no doubt whatsoever that it is not possible to understand the functioning of a complex environment such as a city by looking only at human needs. I'll try to say that more clearly: Studying and planning cities by pursuing only the immediate needs of the people who inhabit them is the best way to ensure that, not long from now, those very needs will no longer be guaranteed.

To understand the physiology of a city, one needs to take into account the entire ecosystem that characterizes it. It is no coincidence that the most significant and respected approach to understanding what a city is and how it functions derives from the work of a botanist: the Scotsman Patrick Geddes. A professor of botany at University College Dundee from 1888 to 1920, Geddes was a singular and eclectic figure who revolutionized our vision of the city and city planning. No small achievement for a botanist. During his academic career, he was both a pioneer of the first systematic studies of the flora of Britain and the theorist of the need to analyze the city by using the tools offered by ecology and sociology in light of the theory of evolution, which he had learned from Charles Darwin.

The centerpiece of Geddes's approach to urban planning is the potent idea that every city is to be considered, in all respects, a living being and the fruit of its history, of its interaction with the environment, of its buildings, and of the social, economic, and ecological networks of which it is composed. Every function of the city, no matter how peculiar, can be assimilated to the internal vital functions of a living organism. So, for example, its roads and rail lines are analogues of the city's arteries, while its communication lines are the nerves by which impulses and ideas move from place to place inside the urban body. For Geddes, even technological innovation is a product of the city; in contemporary terminology, we would say it is an emergent property of the city, rather than the work of humans.[19] Railroads, communication lines, transport, and industry, although materially produced by human activity, are nothing else but the manifestation of the organic model of the city.

Urban planners—the people who plan cities—are concerned with things such as the design of the streets and transportation networks, the portions of territory allocated to residential, industrial, or recreational activities, identifying those models that enable the largest number of people to live together as comfortably as possible in a limited space. The resolution of these problems creates a model that is imposed on the city from outside.

For Geddes, this kind of a priori city planning is bound to fail. Thanks to his training and biological background, he knows that an organic order cannot be human-made. Just as no scientist can create life, no urban planner can create a city. In Geddes's view, reducing the complexity of urban networks to a design, produced by the human mind and therefore limited by definition, is to kill a city. Most of a city's life, in fact, derives from the diversity and multiplicity of the places that humans recognize only as chaos. The various activities that we perceive as confusion—casual encounters with people, simultaneous operations of companies that have no relationship with each other, the thousands of vehicles that move in all directions, the stories and opportunities that intersect and are continuously generated inside of cities—are nothing but components of an organic mechanism that is too complicated for us to understand in its entirety.

Furthermore, Geddes has a clear idea of the importance to a city of all the nonhuman living beings that compose it. During his studies of evolution, he is fascinated by phenomena of symbiosis between different species, which can even lead to the birth of new species with different characteristics. He studies the relationships between plants and animals and gleans from them the strong conviction that the main force that shapes

life is the cooperation between living beings.[20] According
to Geddes, it is Peter Kropotkin's *mutual aid* that shapes
relationships between living beings with greater force, and to a
greater extent, than competition. Species, especially vegetable
species, manage to find a *reciprocal advantage* through the
slow and continuous adjustment of their relationships, guided
generation after generation by evolution. It is thanks to such
a process of coevolution, in which humans, the environment,
buildings, networks, plants, and animals are transformed,
that cities are able to develop and prosper. Consequently, any
attempt at planning must be interactive and based on small
adjustments between places and the inhabitants, human and
nonhuman, of the city. No plan that has as its prospect the idea
of shaping the city will ever be successful because evolution is
open to different solutions that cannot be foreseen. Geddes
believes that, just as with living beings, where phenomena
of aggregation can unite parts or simple organisms to form
complex configurations,[21] so in the urban sphere, aggregations
of simple and casual modules lead to the formation of urban
fabrics and complex configurations.[22]

The revolutionary and still-current work of this Scottish
botanist changes completely the outlook of the urban sciences.
The laws of biological evolution and social organization are
the main engines of life in the city. Thanks to Geddes, the idea
that there exists a metabolism of the city, a physiology all its
own, has become widely accepted today. Only in this way can

it be correctly understood that, like every living being, a city has a constant need of energy and resources in order to grow and that, inevitably, it produces refuse and waste. In order to keep this cycle functioning, the presence of plants inside the urban organism is essential. Unfortunately, from this point of view, the situation is very far from being satisfactory. All you have to do is look at our cities from above to realize that they are completely mineral spaces, built to the last available square foot. The surface area occupied by plants in the city is minimal. In many historic centers that have maintained their medieval or Renaissance design sufficiently intact, plants are totally absent.

It is a situation whose gravity is apparent even to a very superficial analysis, but hard to quantify because of the variety of systems employed to measure the plant presence in cities. A good job has been done by the World Economic Forum, which, working with MIT, has developed the Treepedia program (accessible at sensable.mit.edu). For many of the world's cities, the software measures the percentage of urban surface area covered by arboreal vegetation. Site visitors discover that, among the cities listed, Vancouver, with 25.9 percent of its surface area covered by tree foliage, is the one with the most vegetation, while in many urban centers the arboreal coverage is much less than 10 percent. This is a dramatically low figure if we consider the innumerable advantages for any human activity of having plants in the city. But most of all, it is a figure totally incompatible with the need to diminish global warming, which, it must be recalled, remains the greatest danger to the future of humanity.

I am sure that most of my readers know perfectly well what global warming is and what the causes are that feed it. Nevertheless, since there could still be a few who have some doubts about it, I think it may be best to devote a few lines to a brief explanation of the factors that are changing the climate of our planet. Although the news media sometimes carry misleading reports that might make it appear that there is lack of consensus among scientists about what is happening, in reality there is no doubt about it: The average temperature of the earth is increasing at a rate never experienced before and the main cause of this warming is the increase in the atmosphere of greenhouse gases— mainly carbon dioxide—produced by human activities. I repeat, there is no doubt about this. Rarely in the history of science has there been a phenomenon about whose causes science has reached such a wide consensus.

What we still do not know with precision is what the effects of global warming will be. There exist, obviously, some models that forecast a series of effects tied to the increase in temperature, but no one has a clear idea yet of what could happen when, at the end of the century, if we do not adopt effective countermeasures, the average temperature of the planet will have increased by more than 4°C (7.2°F). Some consequences of global warming are already plainly visible even to the most skeptical eye: record high temperatures; significant increase in violent atmospheric events even in zones of the planet where such events have never happened before; increase in the number and size of fires; rises in sea level; etc. But this does not tell us much about what could really happen in the upcoming years.

The reason for this uncertainty is tied to the fact that temperature is the basis of any physical or biological process. In other words, its importance is such that it is virtually impossible to have detailed models of what will happen. Certainly urban centers—our new ecological niche—will not be left unharmed. On the contrary, many cities are already facing the effects of global warming and, in the future, the situation can only get worse. The reasons are easy to understand. More than 90 percent of cities are coastal and, as such, will be subject to increasingly frequent and dangerous floods due to the inevitable rise in sea level. Increasingly violent atmospheric events—storms, flooding, wind, and drought—will provoke greater damage that, apart from devastating the population, will have important economic effects, causing interruptions in trade and commerce and in the normal operations of cities. Summer heat waves will become increasingly frequent, with disastrous effects on people's health. The increase in temperature will lead to an increase in epidemics and in the number of pathologies.

A study done in 2017 estimates that, even if we manage, before mid-century, to limit the rise in median temperature to only 2°C (3.6°F) compared to the preindustrial level—a prospect that by now is almost unreachable—the number of deaths in cities will be greater than 350 million due to the effects of heat waves alone.[23] As if that were not enough, we have to consider that in cities the effect of higher temperatures is amplified by the peculiar characteristics of the urban environment. Considering only the effects of urban heat islands, that is, the fact that city temperatures are higher than those in surrounding rural areas, it has been calculated that the average temperature increase in cities will be 6.4°C (11.5°F).[24] Such a figure, although highly variable, depending on the geographical location and peculiar characteristics of each urban center, constitutes an indicator of the enormous impact that our way of building has on the environment.

The first to notice this phenomenon was the English chemist and pharmacist Luke Howard, who deserves credit not only for the first observation of the existence of urban heat islands but also for noticing that the temperature difference between the city and the country is greater at night than during the day. In 1820, in his treatise *The Climate of London*—the very first to concern itself with the climate of a city—Howard wrote that based on comparing the data for nine years of temperatures recorded in the center of London and rural areas immediately outside the city, "at night it is 3.7°F (2.1°C) warmer in the city compared to the country."

There are various reasons for this overheating, and they depend on the way our cities are built. One of the main factors in the formation of heat islands is the artificial nature of city surfaces. Because of their impermeability, along with the lack of vegetable coverage, they do not have the possibility to cool through the evapotranspiration of water, contrary to what happens in rural areas. There's more. The dark surfaces in cities absorb significantly greater quantities of solar radiation, and materials such as asphalt and cement have different thermal properties than those typical of surfaces present in rural environments. Moreover, a significant part of the energy used in cities by automobiles and industry, or for heating and cooling buildings, is dispersed in the form of residual heat that increases the temperature of the environment. Finally, the geometry of buildings, the lack of wind that impedes cooling, the greater atmospheric pollution, and the dust particles that change the radiant properties of the atmosphere—everything in the city contributes to raising the temperature of the environment.[25]

If we add together the effect of global warming and the typical warming effect of urban heat islands, the results are not at all reassuring. And, really, how could they be? How can we imagine that global warming has no effect on the health of the places where we live? One of the most urgent problems of the battle against global warming is precisely the challenge of making evident to everyone the consequences of the rise in temperature. Indeed, the gap that exists between the understanding of the phenomenon within the scientific community and the understanding of most citizens is enormous.

The difficulty here is rooted in an inability to find adequate systems of communication. Often, communication stops with the signaling of extreme meteorological events or the enumeration of the risks associated with the warming of the planet (for example, heat stress, worsening of air and water quality, reduced food supplies, increases in the distribution range of tropical-disease carriers, worsening of social conditions, etc.) and produces less than negligible results. The immaterial nature of the indications regarding global warming is not able to transmit the importance of the problem. Most of us are not capable of imagining how a mere two-degree increase in temperature can have such a heavy impact on our daily lives. By themselves, data and facts do not persuade people to change their ideas or behavior. Not at all. It also seems that the acceptance of global warming has nothing to do with people's level of scientific literacy.[26] Rather, it has to do with people's incapacity to create a mental image of the problem. That is why any attempt to enable people to visualize the problem of global warming is to be welcomed.

To make evident through practical examples what the climate of our cities will be like at midcentury—even in the fortunate hypothesis that we manage to limit global warming to the most optimistic levels of today's forecasts—ETH Zurich (the Swiss Federal Institute of Technology) has paired the characteristic climate data from the major cities of the world in 2050 with the present climate in other cities. We thus discover that in thirty years, in the northern hemisphere, both the summers and the winters will be warmer, with respective average increases of $3.5°C$ and $4.7°C$ ($6.3°F$ and $8.5°F$), and that the hemisphere's cities will have the climate that cities six hundred miles farther south have today. The climatic conditions of Rome in 2050 will be similar to today's climate in Izmir, London will be like Barcelona, Stockholm and Oslo like Vienna, Munich like Rome, Moscow like

Sophia, San Francisco like Rabat, Los Angeles like Gaza, Paris like Istanbul, and Madrid like Marrakesh.[27]

While cities are particularly vulnerable to global warming, the good news is they are also the place where global warming can be countered most efficiently. Since cities produce 75 percent of human-origin carbon dioxide, it is in cities that the production can be blocked by using trees to remove the greatest possible quantities of it from the atmosphere.

In 2019, a team of researchers at Zurich Polytechnic published the findings of a study claiming that the planting, on a worldwide level, of one trillion trees was by far the best, most efficient and measurable solution for assimilating from the atmosphere significant percentages of the carbon dioxide produced since the beginning of the industrial revolution.[28] Despite the quality of the study and its solid scientific foundation, criticism was quick to arrive. Where would we ever find the space to plant a trillion trees? How much would it cost? Criticism that was, for the most part, without any basis. The surface area for planting these trees exists and the cost, as high as it is, is much lower than the cost of any imaginable alternative that has even a fraction of the chances of success as this one.

If we then managed to plant a substantial number of these trees inside of our cities, the results, I am certain, would be even greater. Indeed, the efficiency of plants in assimilating carbon dioxide increases in proportion to their vicinity to the source of its production. The closer the plants are to the source of the carbon dioxide, the more of it they absorb. Every surface of our cities should be covered with plants. Not only (the very few) parks, boulevards, flower beds, and other canonical places, but literally every surface, roofs, façades, streets—every place imaginable that can host a plant should have one. Again, the idea that cities have to be impermeable places, made of minerals, contraposed to nature, is only a habit. Nothing prevents a city from being completely covered in plants. No technical or economic obstacles really preclude a choice of this kind. And the benefits would be incalculable. Not only would the plants fix enormous quantities of carbon dioxide where it is produced but the quality of people's lives would be better. From every possible point of view—from the improvement of physical and mental health to the promotion of social interaction, from the reinforcement of powers of concentration to the reduction of crime—plants have a positive influence on the way we live.

Why in the world our cities are not completely covered with plants, despite the thousands of absolutely reliable published studies on the benefits of urban greenery, remains a mystery. After all, that plants improve the quality of the air is something we have known for centuries. In 1661, the writer John Evelyn published a book with the portentous title *Fumifugium or the inconveniencie of the aer and smoak of London dissipated together with some remedies humbly proposed by J.E. esq. to His Sacred Majestie and to the Parliament now assembled*, one of the first studies on atmospheric pollution. It is an extraordinarily modern text divided into three main parts. The first part examines the causes of the bad air quality in London—owing to the fumes provoked by the combustion of coal—and correctly identifies the effects that its inhalation has on the health of the airways. The second part proposes a practical solution to the problem: Coal must be prohibited within the city of London and in its place firewood must be used. Finally, the third part outlines a prospective solution that even today has lost none of its original effectiveness: Improve the quality of the air in London by the widespread planting of trees and shrubs inside and in proximity to the city.

It is a proposal whose revolutionary scope is hard enough to comprehend today but which at that time must have seemed like the idea of an absolute dreamer: fascinating but utterly baseless. Keep in mind that in Evelyn's time nothing was yet known of photosynthesis, whose discovery, thanks to Joseph Priestley and Jan Ingenhousz, would come about only a century after the publication of *Fumifugium*. Any speculation about the capacity of plants to improve air quality was based only on empirical experience, without any theoretical basis. Yet Evelyn, despite the lack of any scientific support and in a time when forests still blanketed large regions of Europe, is an avid supporter of the necessity of plastering our nations with trees.

In 1664, three years after the publication of *Fumifugium*, Evelyn publishes another fantastic volume entitled *Sylva or a Discourse of Forest-trees and the Propagation of Timber in His Majestie's Dominions*, which is the first book published by the newly born Royal Society. It is an encyclopedic study of the cultivation of trees, from the selection of seeds to the needs and adversities faced by all of the most common species that can be grown on British territory. In this regard, it is not the first nor the most original treatise on the cultivation of trees. What makes *Sylva* sensational, however, is that Evelyn uses every means possible to convince his readers of the necessity of planting trees. Any argument is good if it leads to an increase in our awareness of the importance of trees. For example, since the English Civil War, which ended only a few years earlier, had caused considerable deforestation owing to the use of wood for military purposes, Evelyn points to the need to reforest in order to ensure the navy its required supply of wood for shipbuilding.

This is obviously a pretense. Notwithstanding his claim on the frontispiece of *Sylva* that his book is a response to "certain Quaeries propounded" by the "principal Officers and Commissioners of the Navy," many of the species that he recommends planting and that he describes with loving care are not at all suitable to be used as naval material. Evelyn's only interest is the planting of as many trees as possible. The book's Latin quotations on trees taken from works by authors such as Pliny, Horace, and Virgil are there to ingratiate the intellectuals of the era, while the references en passant to famous contemporary tree-loving personalities like the king or "that Lady of Northamptonshire" are there to create the impression that planting trees is a fashionable pastime.

What we need today are thousands of Evelyns, ready to spread the idea that covering our cities with plants could effectively stave off global warming. We need to change our way of representing the city. The image of the urban jungle should conjure up the idea not of a place full of dangers but, on the contrary, of an active participant in the natural environment that, consciously and by way of its trees, contributes to the transformation of our cities into an enduring ecological niche.

A forest is not an ensemble of different trees but a community of individuals. Trees connected by complex root networks exchange everything they need. A magnificent example of mutual aid.

Three

PLANTING THE UNDERGROUND

To understand fully what Patrick Geddes said about the life of the city, we have to return to the main life-shaping force: cooperation among living beings. In nature, this manifests itself in part by way of coalescence, fusion, and grafting. From this perspective, plants offer us, yet again, a model to imitate. They are the masters of mutual aid on the earth, and the story that follows is an admirable example.

Some years ago, during an excursion in a rain forest north of Auckland, two New Zealander researchers, Sebastian Leuzinger and Martin Bader, discovered a stump of a kauri tree (*Agathis australis*) that had some unusual characteristics. For a nonexpert, a stump is a stump, there's not much that's interesting to say about it. Stumps are just the sad remnants of dead trees, destined to be consumed by microorganisms, mushrooms, and forest insects until they disappear. However, in the case of this

particular stump, there was something out of the ordinary.
It was indeed a stump but it was also undoubtedly active.
Although it had no leaves or other organs capable of performing
photosynthesis, its internal tissue was still living. This was
surprising. How could it have remained alive in such conditions?
A tree without leaves, in fact, is not only deprived of its energy
source—not being able to photosynthesize the sugars that allow
it to live—but it is also virtually incapable of absorbing from the
surrounding terrain the water it needs. The transpiration force of
the leaves is the engine that allows the water to be absorbed from
the ground and pumped into the tree. In short, without leaves, a
tree not only is unable to photosynthesize but is also deprived of
its water supply.

So how did that kauri stump manage to go on living? The
two researchers hypothesized that it received what it needed
through its root system, which was directly connected with
the root system of nearby trees thanks to a phenomenon
known as root grafting.[1] Grafting is familiar to most people
as the agronomic technique for fusing two distinct but similar
individual organisms to create a new individual organism formed
from the union of the two bionts (this is the technical term
assigned to the two individuals that participate in the graft).
Normally, a graft is performed with a rootstock, which will
constitute the base (with the roots) of the new plant, and a scion,
which will form the upper part of the plant (with leaves and fruit).
The graft creates a complete vascular connection between the

tissues of the two bionts, which enables circulation from the base to the upper part and vice versa.

The practical possibilities of grafting are countless and that is why humans have been using the technique for thousands of years to create plants that combine the positive characteristics of two bionts. Imagine, just as a simple example, that you have two apple trees, one that is able to grow in arid climates but produces unappetizing apples and another that, on the contrary, is not able to survive droughts but whose apples are magnificent. By joining the two individuals through grafting we can, in theory, produce a new tree that combines the positive characteristics of the two apple trees: a new apple tree that is resistant to drought and produces mouthwatering apples. Generally, the process of grafting is a little more complex, but at least the basics should now be clear. What is not so well known is that grafting also happens naturally, without human intervention. Two similar plants that happen to be in continual contact, with their trunks or branches touching, for example, can in certain conditions fuse into a single individual. It was probably through the imitation of these natural grafting processes that humans first learned to create grafts of their own.

But let's get back to our kauri stump. Just as there exist natural grafts between the upper parts of plants, there also exist root grafts that lead to the fusion of root systems belonging to different plants. Unlike branch and trunk grafts, root grafts are extremely common. A forest, for example, will have grafts between different parts of the same root system (autografts),

grafts between root systems of different trees belonging to the same species (intraspecific grafts), and, finally, grafts between roots systems of trees belonging to different species (interspecific grafts). Known for centuries, but until recently considered nothing more than a botanical curiosity, this phenomenon could revolutionize our idea of the real nature of a vegetable community. That is why, when they found themselves before a living stump, Leuzinger and Bader decided that a careful examination of what they were looking at could amount to something much more useful than the explanation of a mere curiosity.

The existence of, shall we say, "living dead" tree stumps (in homage to the movies of the late George Romero), like the presence of root grafts among forest trees all over the planet, is not exactly breaking news. On August 12, 1833, René Joachim Henri Dutrochet, the famous French physicist, botanist, and physiologist who is best known for his discovery of osmosis, gave a talk at a meeting of the Academy of Science in which he, for the first time, referred to living stumps.[2] Dutrochet reports that he was alerted to the presence of numerous living stumps of white pine (*Pinus picea*) by his brother, a forest inspector, "one of the best-educated men in the service of the forest administration." On the basis of this fraternal alert, Dutrochet went to the forests of Jura in eastern France where he indeed finds that "all the stumps of white pine, remaining from trees cut down many years ago, are full of life, as are their roots." So he starts studying the white pine stumps "which certainly were cut down at least forty-five years earlier, still full of life" and he notices that in the spring the cambium, the generative tissue of the tree's secondary growth, again becomes active.

Dutrochet measures the growth in the width of the stump from the time the tree was cut down and quantifies it as "two centimeters or about eight lines of diameter." This is a unique observation: The secondary growth of the truck, although much slower than normal, has blithely continued for at least forty-five years without any foliage. Dutrochet is very surprised by his findings; what he has seen in the forests of Jura has no explanation. His talk begins, in fact, with the observation that the lymph elaborated, produced, by the foliage of trees is necessary to the survival of the entire plant, roots included. So, then, how had those white pine stumps survived for more than forty-five years? Dutrochet sees no other possibility except to suppose that the roots of these stumps are also able to produce small quantities of elaborated lymph, sugars. It is the only incorrect conclusion of his splendid study, as we shall see.

Our knowledge of root grafting, both in angiosperms and in conifers, is not new either. We have known about this phenomenon for more than a century. In the mid-1960s, Barry Graham and Frederick Herbert Bormann produce a list of over a hundred and fifty tree species for which there is scientific evidence of root grafting.[3]

The trees create underground connections, and that these are probably capable of keeping stumps alive for decades is acquired knowledge, albeit always considered a simple botanical curiosity, an oddity found among plants without any value on a more general level. Wrong. The story told by these living dead stumps and the underground communications between distant trees is something so new and fascinating as to change our very conception of what a tree is.

Regardless of the differences among species, trees have always been considered isolated individuals. Sure, not individuals in the animal sense of "indivisible" but indisputably single living organisms with needs and behaviors distinct from those of their fellow trees. In this respect, trees are no different from animals. But is this really true? Hundreds of research studies conducted over the last twenty years seem to be demonstrating a completely different reality. Not single trees but rather enormous connected communities, which through their root systems are able to exchange nutrients, water, and information. Extended communities that sometimes even include plants of different species and that depend for their survival more on cooperation than on competition. A true conceptual revolution whose consequences are not easy to predict.

Nevertheless, this does not help us make any significant progress in resolving the problem of the zombie stumps. Indeed, while numerous recent scientific articles of the underground connections of trees—whether direct by way of root grafting or mediated by fungus networks—have unequivocally demonstrated the advantages of belonging to these communities, in the case of stumps the advantages are not so clear. Why should the surrounding trees keep stumps alive? And above all, how can this exchange of water and nutrients take place in a stump which, lacking foliage, also lacks the driving force needed to move the liquids inside its own body? This last question is what really puzzles our New Zealanders as they are observing the fascinating zombie stump they have happened upon. Unlike the many others who had examined the same phenomenon in the past, Leuzinger and Bader are not content to observe and describe it. They want an explanation for this apparently paradoxical phenomenon. They carefully mark the exact spot where they found their zombie stump and go back to the university in Auckland to gather their thoughts and decide how to proceed. The possibilities are numerous but their resources are limited. So they decide to concentrate on the measurement of the cycles of water absorption inside the stump and the neighboring trees. Their hope is to be able to observe a synchrony in the behavior of the flows or of anything else that might suggest an effective and functioning underground hydraulic connection.

So having assembled the equipment and instruments needed for the study they have in mind, they go back to the forest and begin installing on the stump and the neighboring kauri trees sensors that are able to record in real time the water flow inside the trunk. They set up all the instruments and, having assured themselves that everything is working as it should, go back to Auckland where, seated comfortably in their laboratory, they can follow, thanks to the Internet, the constant flow of data coming from the forest.

After the few weeks necessary for gathering a sufficient number of day-night cycles, they are ready to analyze the data.

What they had suspected from the very first moment they came upon their stump happily takes shape right before their eyes. The cycles between the closest kauri tree and the stump were without a doubt inversely related. When the flow increased inside the tree trunk, it diminished in the stump and vice versa. But what seemed even more interesting was the opposite behavior of the stump with respect to the whole tree.

As we know by now, for water to be absorbed from the ground and transported to the foliage it is necessary for the leaves to lose water through a process called transpiration. This process requires the stomas—the thousands of tiny valves on the surface of every leaf—to open, releasing the water vapor inside the plant into the surrounding atmosphere. It is precisely this water loss, provoking a depression in the plant's closed vascular system, that causes the upward pull from below and the absorption of the water from the ground. Since the stomas (whose opening sets off the whole process) open during the day in response to sunlight, the peak of water absorption inside the tree occurs during the daylight hours.

This is the *standard* behavior that the two researchers observe in the water flows inside the intact tree. The new and interesting thing that emerges from their measurements concerns the hydraulic behavior of the stump. On very sunny days, in which the tree loses lots of water through its leaves (transpiration), there was a prodigious transport of water in the nearby trees, while the stump saw no activity at all. At night, on the contrary, in the absence of transpiration in the neighboring trees, the hydraulic flow in the stump rose to its maximum level. On days with little sunlight or on days following intense rain—days of little or no transpiration in the neighboring intact trees—the water flow in the stump remained high even during daylight hours. In short, everything came together to demonstrate a close hydraulic relationship between the neighboring trees and the zombie stump. There remained one last thing to explain. What made the water rise inside the stump? Here, there is no data, but the most plausible explanation seems to be a movement brought about by osmotic force. If that is actually the case, our story will have started with Dutrochet, who was the first to describe these living stumps, only to return a hundred and eighty years later to the same Dutrochet, the discoverer of the phenomenon of osmosis. The flux and reflux of science.

Having clarified the mechanism, the most important question still awaits an answer: Why? Why should healthy trees support a stump for decades? This fact seems totally incomprehensible. We are accustomed, in fact, to the idea that competition and the struggle for survival are the driving force of evolution. For more than a century, starting with the revolutionary work of Charles Darwin, the prevailing idea that has forged our notion of the functioning of the community of the living has been that the driving force of evolution is competition, the survival of the fittest. Nevertheless, right from the very beginning, equally authoritative voices were raised against those who, declaring themselves the heirs and custodians of Darwinian thought, succeeded in imposing the idea of competition as the dominant and regulatory force in relationships among living organisms. I have in mind the voice of the unforgettable Prince Kropotkin, a spokesman for the idea that cooperation or, as he poetically called it, "mutual aid" or "mutual support" is the keystone of the entire history of evolution.

Although today the amount of evidence in support of the fundamental role of cooperation in the evolution of living species has been enormously enriched, the idea continues to be perceived as marginal compared to the solidity of its competitive counterpart. Why? I am convinced that the main reason for the scarce interest in the study of cooperation as an evolutionary force is tied to the fact that most—the near totality—of evidence in support of this theory comes from the realm of plants, which as such are considered unimportant. The anthropocentrism or, if we want to be magnanimous, the animal centrism that afflicts the world of science is a serious problem. Our vision of the world as a place in which conflicts and privation are the basic forces that dominate evolution is a classic example of this animal distortion. Well-known mathematical models, such as the model of interspecific competition—the predator-prey model—developed by Vito Volterra and Alfred Lotka in 1926, though regarded today as universally valid, were originally developed to describe relationships among animals.

Not long after the end of the First World War, Umberto D'Ancona, one of the most important Italian zoologists of the twentieth century, noted while studying the fish populations in the Adriatic Sea that the percentages of the different species caught in the nets tended to fluctuate regularly. In an effort to understand this fluctuation, D'Ancona consulted Vito Volterra, an important mathematician who would later become his father-in-law, and Volterra developed the model that explained it. Apart from the beauty and the indisputable validity of the predator-prey model, the fact remains that this model, which has profoundly influenced the study of the dynamics of natural populations, and more generally, I would say, our idea of the relationships among species, is a model conceived, elaborated, and tested in response to typical needs of animals. What does the predator-prey model have to do with the world of plants? It is not the only case. Many models which have had a great influence on our awareness of the functioning of communities, perhaps not so well known by the general public but very influential among experts in the field, are valid almost solely in the animal kingdom and absolutely cannot be considered generally applicable.

I would like the absurdity of this proposition to be clear. Discoveries made in the vegetable kingdom are not thought deserving of any attention unless and until they are replicated in the animal kingdom. Conversely, models whose validity is obviously restricted to the animal world are, ipso facto, thought to be of universal value. Think of the irrationality of this position: The discoveries made in 85 percent of living beings (plants) are required, in order to be held universally valid, to be confirmed in the 0.3 percent of living beings that belong to the animal kingdom. And so we go on living with the ridiculous and dangerous idea that what is valid for the noble 0.3 percent of life (animals) is what characterizes all of life and is worthy of being known; all the rest is marginal. I wonder if the irrationality of this situation strikes you as much as it does me. That the 85 percent of life represented by the vegetable world is, by itself, the only and indisputable representative of life on our planet, is of no interest to anyone. It is as though in order to become a law, a bill proposed by 85 percent of the representatives in a parliament were required to be reviewed by just 0.3 percent of the same parliament who, at their discretion, may approve or reject it.

Let's try to transform these impersonal percentages into absolute numbers by applying them to the same parliamentary example. If we imagine a parliament made up of 500 elected representatives—a number in line with the average European parliament—the 85 percent of plant life would have 425 representatives while the 0.3 percent of animals would have 1.5. The rest would be assigned to fungi and various microorganisms. So we would have 1.5 (let's round up to 2) representatives who decide for everyone. When something like this happens in our parliaments we call it dictatorship. This fundamental misunderstanding of what life really is and how it functions is one of the most insurmountable problems of modern times. Please do not underestimate its importance. As long as we hold on to this mistaken idea of our position among the living, the very survival of our species will rest on an insecure basis. The fact remains that ecological models based on animal competition have become the models with which we describe the world of the living, while models based on a cooperative vision of life, because they come mostly from the vegetable kingdom, are ignored. And this despite their having, in some cases, the force and support of such a huge quantity of evidence that it would seem impossible not to take them into account.

So let's go back to the question of questions: Why should healthy trees burden themselves for decades with the care of a neighboring stump? At first glance it seems like an absurd response, a pure and simple waste of resources. And all living organisms, except for humans, tend to avoid waste. So? So there must be an explanation and it must necessarily result in some advantage for the intact trees.

One way to arrive at the explanation could be to evaluate, first of all, what the advantages are for a community of trees in remaining connected through root grafting and, once we have identified these advantages, determine whether those advantages persist in the presence of a stump. Unfortunately, although there are hundreds of publications that describe the phenomenon, only a few raise the question of the possible evolutionary advantage for the trees.

Why unite? It reminds me of the question Darwin posed when the time came for him to decide to get married or not.[4] On the face of it, it's almost impossible to see any advantages to uniting in matrimony, but in the end, nearly all of us, including Darwin, opt for stable relationships. The subterranean connection of plants resembles this dilemma a lot and we are finally beginning to understand its advantages. Let's have a look at some of them and, so as not to veer too brusquely from the issue of matrimony, let's say right away that a neighbor in good health with whom you can cooperate by way of root grafting, instead of competing, can be a partner with whom to reproduce.

Furthermore, a neighbor in good health has less probability of becoming a source of disease. For stationary beings like trees, not having individuals in bad health nearby, potential sources of contagion, has much greater value than it does for the animal kingdom. What else? By way of a root graft, a tree could acquire from its neighbors beneficial fungi or microorganisms. Obviously, the risk of transmittal of pathogens by way of the same mechanism should envisage the ability of trees to distinguish between healthy and unhealthy neighbors.[5] By the same technique, since roots produce many toxins that are useful for the plant's defense, trees might enrich their arsenal of chemical weapons against various pathogens and herbivorous insects. Finally, the most evident advantage of all: the increase in the stability of trees.

In 1988, Jon E. Keeley, working on plants of *Nyssa sylvatica*—a midsize tree whose origin in swamp lands is highlighted by its name, borrowed from a naiad (water nymph)—noticed that they tended to form many more root grafts compared to tree populations originating in other habitats. Since the anchorage of trees in swampy environments is always problematic, root grafting with other trees significantly increases the trees' mechanical support.[6] The same principle is valid in windy environments in which trees are subject to similar problems of instability. Support for this hypothesis also comes from the behavior of large trees, that is, trees with more risk of instability, which tend to form greater numbers of root grafts, as well as from the higher resistance of trees connected by root grafts to extreme atmospheric events.[7]

Now that we have a clearer idea of what some of the advantages might be of uniting one's own roots with those of neighboring trees, let's see what happens if instead of a whole neighbor, we have a stump. Paradoxically, in some respects the advantages of being connected to a stump may be even greater. Think, for example, of a root graft between two healthy trees, in which due to some event one of the two is struck down. The surviving tree will suddenly find itself with a double root network at its complete disposal with all the advantages that come with it, not least the greater stability that an extended root network anchored to those of nearby trees can provide. That is not something of little consequence, especially in times like ours. Indeed, it is worth recalling that, because of global warming, the number of extreme atmospheric events is increasingly worrisome everywhere, as is the number of fallen trees.

Red spruce are the trees of music. Antonio Stradivari made fourteen violas and violins starting from a single red spruce.

Four

PLANTING MUSIC

W ind is the main adversary trees have to defend against, at least in Europe. More than 50 percent of the damage suffered by European forests is caused by wind. It is not fires (16 percent of damage) or pathogens, or insects that threaten our forests, but wind. From 1950 to the present, wind damage to European forests has been increasing steadily. From 1970 to 2010, the number of lost trees doubled, going from about 50 million to 100 million cubic meters.[1] This is an enormous number of trees that, by falling to the ground, cause radical change and instability for the landscape and the ecosystem. They also reduce the capacity of the damaged areas to absorb carbon dioxide by 30 percent.[2] Moreover, the quantity of carbon dioxide in the atmosphere is the primary cause of global warming which, in turn, is one of the reasons why we are experiencing an increase in the frequency and intensity of such destructive events.

From October 28 to 30, 2018, a tempest of wind and rain struck ample regions of the eastern Alps with a wind velocity exceeding 120 miles an hour. An astounding number of trees were blown down and tens of thousands of acres of forest were wiped out. The consequences of this natural disaster went well beyond the direct damage to the forests, setting off a series of circumstances that nobody could have foreseen. The worst of these, in my view, was the damage done to a forest of red spruce, whose wood had provided the soundboards for great musical instruments for centuries. Causes and effects: because of the carbon dioxide dispersed in the atmosphere the temperature of the planet rises, atmospheric phenomena become violent, and windstorms destroy the trees from which great violins have been made for hundreds of years.

The way I see it, musical instruments are the most wonderful things that people have ever thought of making out of wood, and Antonio Stradivari is the luthier who achieved the greatest success in the miracle of transforming this material into celestial sound.

Stradivari is the end result, the apex, of the school of Cremona, the world's most important school of violin makers in terms of the innovation and quality of its instruments. Cremona is not only Stradivari. Other violin makers able to infuse their instruments with the breath of immortality also worked and headed workshops there. The now legendary

luthiers of Cremona include Nicolò Amati (1596–1684), the
progenitor of the Amati family, who is considered the pinnacle
of sixteenth-century violin makers, and Giuseppe Guarneri,
also known as del Gesù (1698–1744).

You have no doubt heard it said that it is impossible to re-create
the sound of these instruments. Although their construction
techniques, woods, and varnishes have been analyzed in every
possible detail by the best-equipped laboratories in the world, no
one has ever managed to unlock the secret of these extraordinary
artists of sound. Ask an accomplished violinist to play a modern
instrument constructed with the most maniacal accuracy and
then hand that musician a Stradivarius: The sound will be fuller,
more powerful, the notes more distinct. The bass notes and high
notes will fly off the bow with the same incredibly expressive
clarity. Someone who has played a Stradivarius knows that from
that moment on they will never feel the same sensation with any
other instrument.

Don't believe anyone who tries to tell you that the results of
a "blind" test have shown that expert musicians preferred the
sound of another violin or that they were not able to recognize
the sound of the Stradivarius. It's not true. Either they were,
what I would call, inattentive musicians or the Stradivarius
had undergone so many and such extensive modifications that
it no longer had anything, or certainly very little, left of the
original instrument. Many Stradivarius violins, either because
of accidental damage or simply out of pure ignorance, have been
substantially modified. In such cases, the famous cartouche

Antonius Stradivarius Cremonensis Faciebat, with an indication of the year of construction, are testimony only to the memory of an instrument that no longer exists.

Instruments still conserved from the golden years of Stradivari, from 1700 to 1725, are beyond compare, instruments so famous and imbued with history as to be known by name. Violins such as the Messiah, considered along with the Lady Blunt (from the name of one of its owners, Lady Anne Blunt, a granddaughter of Lord Byron) to be one of the best-conserved Stradivariuses. Or the Duport violoncello, named after the musician Jean-Pierre Duport, which has a small dent in its ribs that legend has attributed to the boots of Napoleon, inflicted during his imprudent attempt to play it. And then there are the Cremonese, the Vesuvius, and the Maréchal Berthier violins, and the Macdonald viola, all instruments whose extraordinary quality is enhanced by the fascination of the historical vicissitudes they have gone through in making their way down to our own times. Getting to know the history of the Stradivarius instruments is a journey through a large part of the last three centuries of Western civilization.

Perhaps only Giuseppe Guarneri, the creator of the famous Cannone, the violin of Niccolò Paganini, reached the point of producing instruments endowed with the same force and charm. But where does this magic come from? The search for an answer to this question has involved generations of musicians, luthiers, scientists, and experts in wood and materials, each of them convinced they have finally found the solution to the mystery in some detail or other: the necks, the varnishes, the seasoning, or the treatment of the wood. The truth is that no one has ever succeeded in discovering the real secret.

Unless it is, obviously, the red spruce. Not that it is the only wood used by luthiers in making violins. There are many others: mountain maple for the back and sides, ebony for the fingerboard, rosewood, and boxwood are just some examples. But the red spruce is the heart of it. It is only with red spruce that one can make a soundboard worthy of Stradivari. For the most part, the quality of the instrument depends on the quality of the wood used to make the soundboard. And the right red spruce, the one that produces so-called *resonant wood*, does not grow everywhere but only in a few select forests. For his soundboards, for example, Stradivari used only the resonant wood produced from red spruce in the Paneveggio forest, in the Italian Alpine region of Trentino.

But what is it that makes red spruce (*Picea abies*) the perfect tree for the creation of soundboards?[3] First of all, its flawless conduction of sound ensured by, according to luthiers, its minuscule resiniferous canals, which run along the entire length of the trunk and which, after seasoning, remain hollow, allowing air to vibrate inside them as if they were microscopic organ pipes. The crystallization of the resin on the walls of these canals, which comes about only after a long natural seasoning, is said to be fundamental for the further improvement of the wood's capacity to transport sound.

To attain high-quality resonant wood, the trees must have a diameter of at least two feet, dimensions that are reached generally after a hundred and fifty to two hundred years. That's not all. In order to attain the right density of the wood—necessary both to transmitting sound and to holding up under the extraordinary mechanical stress to which a violin soundboard (one-tenth of an inch thick) is subjected—the trees must grow to considerable heights, preferably on northern slopes and in poor terrain. The tree's growth must be slow and regular; its trunk must not be twisted, knotted, or otherwise altered; and it must be cut down during a period of vegetative repose, when the sugars in the tree have been transformed into starch and the wood is more resistant. Finally, the cut. It must be done in "quarters," and not in boards, so that the surface of the wood remains as perpendicular as possible to the tree's concentric rings. The truth is that only very few trees are really suitable for making prestigious soundboards.

The climactic conditions during the tree's growth are a fundamental factor. Spruce trees must grow in an environment that is both stable over the years and not too favorable to growth. Indeed, the annual growth rings must be as regular as possible in order to produce wood that is uniform. One of the secrets of the great instrument making of the seventeenth century seems to lie in the particularly slow growth of spruce during this period, owing to the advent of the so-called Little Ice Age.[4] This particularly cold period, which struck Europe between the fifteenth and nineteenth centuries, had its coldest temperatures in correspondence to the seventy years of the Maunder Minimum, a period of reduced solar activity named for the English astronomer Edward Walter Maunder. When you examine the growth rings of trees that developed in Europe during those years, you notice how small the distance is between one growth ring and the next. The Little Ice Age, then, could be one of the many factors resulting in the quality of the Stradivari violins. If this were so, it could be increasingly difficult in the future to find wood with the same excellent characteristics. On the one hand, global warming provokes the storms that knock down the forests which produce resonant wood, while on the other hand, it alters the characteristics of the trees' growth.

Some years ago, I had the opportunity to examine up close some of these exceptional musical instruments. Along with my friend and colleague Marco Fioravanti, a great expert of woods used in making works of art (paintings, sculptures, furniture, and, of course, musical instruments), and two of my coworkers, Elisa Azzarello and Cosimo Taiti, I traveled to one of the world's holy temples of stringed instruments: the Violin Museum in Cremona. The opportunity offered to us was unique: to study some of the violins, violas, and violoncellos made by masters such as the Amati family, Antonio Stradivari, and del Gesù for the purpose of analyzing the volatile substances produced by the wood parts of the these instruments.

We had some hypotheses about what happens to wood as it ages in particular artifacts like stringed instruments, and in order to test them, we had to analyze some old instruments. To substantiate our laboratory-tested hypotheses with real data, we had requested permission from the Violin Museum to have access to its treasures. Through Marco, who was a long-standing collaborator of the museum, we received notice of the availability for our analyses of many of the instruments in the collection. For twenty-four hours, on closing day for the museum, we would be allowed to analyze some of the all-time wonders of violin making.

As you can certainly imagine, working with these instruments is not at all a simple proposition and, above all, is subject to a thousand precautions and restrictions. Among them, naturally, the absolute impossibility to remove any, even microscopic, part of an instrument. The instruments are untouchable. You can study them only insofar as the investigation does not require any interaction with the materials of which they are composed. In response to these restrictions, our study was to be based exclusively on the analysis of the volatile substances emitted by the wood. The design of our experiments required weeks of study and preparation. We brought to Cremona from our laboratory in Florence everything needed to set up a modern laboratory for the analysis of volatile compounds inside the room assigned to us in the museum. To place our sensors inside the sound boxes, the idea was to utilize the only apertures present on the instruments: the so-called f-holes and the buttonholes on their ends. By closing all the apertures for timed intervals, we would have a way to find out how many molecules were produced in that amount of time by the wood of the violin.

We got to Cremona two days before the fateful date and prepared the laboratory that was to host our illustrious patients. During those two days, we got to know the museum and its fantastic collections. I still remember our first visit, guided by the directress. Everywhere I turned, wonderful musical instruments from different periods, ancient and modern, were on display with the marvels of their engravings and decorations or the austere simplicity of their formal perfection, reminding us of the importance of the Cremonese school of violin making. Proceeding from the origins of the violin to a luthier's workshop, from the history of the school in Cremona to the Stradivarian curios, we finally arrived at the heart of the museum, the famous Room 5. The name by which it is known, "the treasure trove," renders only partially the idea of what it contains. It is as though you are entering the sanctum sanctorum of Cremonese instrument making. On magnificent display, each in its own glass case, nine masterpieces are offered for the adoration of the planet's musicians. The directress describes them for us one by one with the familiarity and affection of a parent for her children: "This room contains an exhibit of some of the masterworks of the Amati family. One for each illustrious member of the family. Here before us the Charles IX by Andrea Amati, made by the progenitor of the family around 1566; next to it the Stauffer violin from 1615, the work of Girolamo Amati, one of Andrea's sons; and finally, the splendid Hammerie violin from 1658 by Girolamo's son, Nicolò Amati. Before us now, the museum's three Stradivarius violins: the Clisbee from 1669, the celebrated Cremonese from 1715, and the Vesuvius from 1727; over there one of his violoncellos, the Stauffer ex Cristiani, from 1700. And finally, the Guarneri: the Quarestani violin by Giuseppe Giovanni Battista Guarneri, made in 1689, and the *Stauffer* from 1734, by his son Bartolomeo Giuseppe Guarneri del Gesù."

The next day, at dawn, we were set to go, dressed in our white coats, waiting for the arrival of the longed-for masterpieces. The first to arrive, noblesse oblige, was Andrea Amati's Charles IX. I touched it with reverence, as though I were handling something sacred. That violin is from 1566! It was played for four hundred and fifty years, created by the hands of Andrea Amati for Charles IX of Valois, I mean, Charles IX! Son of Catherine de' Medici, the king of the Saint Bartholomew's Day massacre, of the Amboise conspiracy, the Duke of Guise, the Prince of Condé. The unfortunate Charles IX. To remind myself of what he looked like, I had gone to see his portraits painted by François Clouet: Charles IX, age eleven, in 1561, at the age at which he became King of France, and then in 1572, two years before he died of tuberculosis, when he was only twenty-four. In the two portraits, the first of a little boy and the second of a young man, the same melancholy and serious look in his eyes. The chronicles recount that he had a passion for music. Maybe it was to make him smile, I thought, that in 1565 Catherine decided to buy him thirty-eight instruments (twelve small violins, twelve large violins, six tenors, and eight basses). Since they were meant for a king, she could not but turn to the best luthier of them all: Andrea Amati.

Unfortunately, only of a very few of those instruments have survived. During the French Revolution, nearly the entire collection was destroyed or lost. The violin in my hands was one of the few survivors from that royal order. It was a magnificent violin; on its sides decorations in gold leaf with the motto of Charles IX, *Pietate et Justitia*, and on the back the crest of the King of France between the figures of Piety and Justice. But quite apart from the richness of the decorations, it is the grace of the instrument itself that is striking, the elegance of the curl, its proportions. The fact that it has produced celestial music for nearly half a millennium. How I wished I knew how to play the violin. Much more prosaically, we limited ourselves, under the attentive gaze of the museum conservator, to measuring the volatile compounds emitted by its sound box. On the red velvet cloth we had prepared for that day, there passed in review before us the history of stringed instruments. In rapid succession, we had the chance to analyze, along with the Charles IX, the entire contents of the "treasure trove" of the museum of Cremona: the Clisbee, the Cremonese, the Vesuvius, the Stauffer ex Cristiani violoncello, the Quarestani, and many other wonderful instruments passed right under our noses, each with its sequence of well-heeled owners, legendary musicians who had played it, dents and scratches, wars, thefts and miraculous recoveries, little adventures and History with a capital H. For one whole day, we joined our names to those instruments and became part of their experience. Regardless of any scientific findings we obtained, we couldn't have hoped for anything better.

If you were expecting some sensational revelations about the mystery of their sound, please know that in the end the only secret of these jewels resides in the quality of their raw material. That is, in the choice of the wood used to make them. In a certain sense, the art of the luthier is all in the choice of the right tree. An incredibly difficult choice that requires long experience, an eye trained to note the most minuscule differences, the intuition of a genius, and certainly a healthy dose of luck. When a tree is found with all the right characteristics, it could wind up being used over the years by different luthiers, as we know happened in the case of a red spruce whose wood was used to produce at least three violins: one in 1744 by del Gesù in Cremona, one in 1746 by Sanctus Seraphin in Venice, and the last one in 1767 by José Contreras in Madrid.

Are you surprised? Thanks to a study performed on thirteen violins made by Andrea Amati, we know that five of them came from the same tree. But the record, yet again, belongs to the greatest of them all, Antonio Stradivari. From 1695 to 1705, he made at least fourteen violas and violins using wood from the same red spruce.[5] In short, these instruments are *wood brothers*! Sometimes the father luthiers of these brothers are different, but much more often they are the same. When a master luthier found a tree that had the highest degree of all the required characteristics, he guarded it jealously and did everything possible not to share it with others. The wood of the red spruce, with its homogenous, compact structure and its regular growth rings, is the fundamental material with which incomparable masterpieces have been made over the centuries. The same growth rings that, as we shall see, allow us to study the climate, date archaeological discoveries and analyze solar cycles.

Adansonia digitata. The African baobab is the most widely dispersed of the species *Adansonia*. Its longevity is legendary.

Five

PLANTING TIME

O f all the stories about plants that are, like music for Edward Elgar, "in the air," none exemplifies better the thin green line that runs, inexorably, through the history of science than the story I am about to tell you. It is a fantastic adventure that ties together tree rings, sunspots, and archaeological dating of the earliest human civilizations. Botanists, Nobel Prize winners, astronomers, archaeologists, visionaries, and adventurers are the unlikely leading players in this saga that, initiated by a colossal mistake, concluded with one of the greatest successes of modern science. And it all starts, yet again, from the concentric circles of trees. But let's take it one step at a time.

We all know that trees signal the so-called secondary growth of their trunks by producing each year a new ring that marks the increase in their circumference. What perhaps we do not all know is that by studying the width of these rings it is possible to understand the climate trends of past seasons. The first to have this amazing intuition was Leonardo da Vinci, who noted: "The circles in the branches of sawn trees, show the number of their years and which years were wetter or drier, according to their greater or lesser thickness."[1]

An extraordinary insight, so far ahead of its time that, like many of Leonardo's discoveries, it would remain a dead letter for centuries. Indeed, it would not be until the early years of the twentieth century that growth rings would be fished out of the oblivion into which they had fallen. Andrew Ellicott Douglass, an astronomer born in 1867 in Windsor, Vermont, brought them back onto the stage as the leading actors in one of the most important scientific dramas of the last century. You might well ask why an astronomer would be interested in the growth rings of trees. Well, that's exactly what makes research so wonderful. You never know where the solution to the enigma you're studying is going to come from. That is why the foremost gifts of a scientist are insatiable curiosity and an unquenchable thirst for all kinds of knowledge.

In Douglass's case, his torment was the rhythm of solar magnetic activity, or the cycle that regulates the appearance of spots on the surface of the sun. These spots are not constant but vary in number and intensity over a period of about eleven years. Galileo was the first to observe the performance of the solar cycle with his telescope in 1610, but it was not until 1849 that regular recordings of this phenomenon were made by the Zurich observatory. It is immediately obvious that if you are trying to study a star that has been in existence for 4.5 billion years, having at your disposal only one series of observations limited to the year 1610 is a fairly big obstacle. Any hypothesis based on such a limited set of data has got to be shaky. This was the problem that was so agonizing for Douglass: how to find reliable data on the sun's magnetic activity prior to Galileo's first observations in 1610.

Douglass began his career as an astronomer in 1894 at the observatory of Percival Lowell in Flagstaff, Arizona. Here, he specialized in the observation of the so-called "Martian canals"— some presumed structures (it was not yet clear if they were natural or artificial) identified by Giovanni Virginio Schiaparelli, the director of the Brera astronomical observatory in Milan—on the surface of the planet Mars during the planet's great opposition in 1877.[2] Schiaparelli's findings—spread over three publications with the not very imaginative titles *The Planet Mars* (1893), *Life on the Planet Mars* (1895), and a further *The Planet Mars* (1909)—had a strong impact on the general public, giving rise to countless theories regarding the presence of sentient life on the red planet.

After some initial incredulity, Lowell also became charmed by the hypothesis that the canals on Mars were actually enormous hydraulic structures built by Martians to manage the scarce water resources on their planet. It didn't take long, in fact, before Lovell became the biggest supporter and popularizer of the theory in the English-speaking world, and not to be outdone by Schiaparelli, he, too, published three books on the subject—*Mars* (1895), *Mars and Its Canals* (1906), and *Mars as the Abode of Life* (1908). Lowell's books made a decisive contribution to the spread of the long-lived legend that Mars had once hosted forms of intelligent life. Most of the scientific community, however, was in substantial disagreement with the idea that there were even canals on Mars—whether natural or artificial—believing instead, as proposed by the astronomer Vincenzo Cerulli, that they were nothing more than optical illusions.

To respond to these criticisms and prove the actual existence of the canals, Lowell had built an advanced astronomical observatory in Flagstaff. And here, at the court of Percival Lowell, is where we find the young Andrew Ellicott Douglass intent on observing Mars and its phantom canals.

Starting one's research activity by looking for evidence in support of such a crackpot theory is not the best way of furthering your career in the world of astronomy. So, after having photographed, between 1894 and 1901, more images of Mars than any other living astronomer and having become convinced, to the ire of Lowell, that there was not so much as one canal on Mars, Douglass shifts his attention to the question of sunspots, dreaming up a system that could significantly extend the minuscule series of observations begun by Galileo in 1610. Douglass wants a series that goes back at least to prehistoric times, and his idea for obtaining it, though simple, is extraordinarily powerful.

Since the number of sunspots is directly related to the amount of solar activity, such that the greater the activity of the sun the more spots are produced, then, he concludes, it can reasonably be assumed that the greater amount of solar energy that strikes the planet during periods of maximum solar activity influences the earth's climate. Furthermore, the climate directly influences the width of the concentric rings of trees, such that years climatically favorable to plant growth induce the production of thicker rings, while climatically unfavorable years induce thinner rings. Therefore, it should be possible to find the famous eleven-year cycles of solar activity in the sequence of thicker and thinner rings on the inside of tree trunks.

According to Douglass, this hypothesis could be confirmed for the years from 1610 to today for which we have both the sequence of solar activity and still-living trees that can provide us with their sequence of rings, what Douglass called "chronology." If everything works as expected, the chronologies obtained from still-living trees could be extended back to prehistoric times by superimposing them on chronologies obtained from dead trees or from wood used to build ancient buildings or structures. It is undoubtedly a fascinating idea, whose validity, however, depends on confirming the hypothesis, at the time completely unsubstantiated, that it is possible to see the solar cycle in the chronology of trees.

In 1906, Douglass leaves Lowell's observatory and becomes a professor of physics and astronomy at the University of Arizona at Tucson. Here he continues his program of research, developing a system he calls "cyclogram analysis," used to identify the periodicity in the series of growth rings. His findings appear to confirm that the ring sequences follow a cycle of eleven and a half years. But it is only a false impression. Unfortunately, cyclogram analysis also yields, in addition to the eleven and a half year cycle, cycles of twenty-three, twenty, nineteen, and fourteen years, and many other shorter cycles as well. In short, despite all of Douglass's claims that it is possible to identify with precision the eleven and a half year cycles caused by solar activity, the system doesn't work.

Evidently, the influence of solar activity on the climate of the planet, assuming it exists, is so small that it does not induce modifications in the width of growth rings or, alternatively, the growth modifications induced by solar activity are masked by much stronger influences—for example, the availability of water for the plant. In any case, the idea of being able to use the chronology of trees to find the cyclicality of solar activity turns out to be irremediably wrong. For everyone . . . except for Douglass, who despite the mountain of negative findings remained relentlessly committed to the soundness of the idea and convinced that, in the long run, by increasing the number of samples, the cyclicality of solar activity will be made clear by arboreal chronology.

It was to our great good fortune that Douglass was so stubborn. Indeed, if he had been more reasonable, he would not have had any interest in constructing longer chronologies that went all the way back to prehistoric times, and the serendipitous consequences of his research would not have manifested themselves, or they would have been delayed for who knows how long. On the contrary, completely unfazed by his discouraging findings, Douglass devotes himself to the elaboration of his first long chronological series, the one for the Colorado Plateau, a desert region in the American Southwest. The high plateau, which is located (with a certain degree of approximation) in the Four Corners, the only point in the United States where the borders of four states—Arizona, Colorado, New Mexico, and Utah—meet, is a vast region that lends itself perfectly to the creation of the first dendrochronological series in history.

The desert climate preserves dead trees from rot indefinitely, enabling them to provide precious chronological series. Moreover, the high plateau region is rich in ruins left by ancient human settlements. Here, ancient trunks and wood artifacts can be found along with a very wide range of tree species, including the yellow pine (*Pinus ponderosa*) and the Douglas fir (*Pseudotsuga menziesii*), which turn out to be especially suitable for dendrochronological measurements in that they form rings that are very clear and sufficiently sensitive to annual climatic variations.

It was only a question of time before Douglass, perennially caught up in the search for ancient pieces of wood coming from the Colorado Plateau, crossed paths with the archaeologists interested in dating the ancient ruins. And so it went. In 1909, after reading an article by Douglass on the relationship between climate and growth rings in trees, Clark Wissler, of the American Museum of Natural History, decided to write to him: "Your work suggests to me a possible help in the archaeological investigation of the Southwest... We do not know how old these ruins are but I should be glad to have an opinion from you as to whether it might be possible to connect up with your modern and dated tree specimens [wood specimens] from these ruins by correlating the curves of growth."[3]

So began Douglass's interest in the dating of the ruins present in the American Southwest. In the absence of a continuous chronology for the region, the first step was to gather the sequences present in the beams used for the construction of the principal archaeological sites and analyze them to identify any correspondences. This approach would not lead to an absolute dating but would provide, at the very least, an extremely interesting relative dating. In brief, Douglass would be able to establish which sites were built in the same year and which were built before or after that year, without, however, being able to give the exact years of their construction.

In May 1919, ten years after receiving the letter from Wissler, Douglass presented his first extraordinary findings from his study of growth rings. After examining samples from Aztec ruins in New Mexico and comparing them to others from another site, Pueblo Bonito, also in New Mexico, Douglass could state with absolute certainty that the second site had been built forty years earlier than the Aztec site. No one had ever been able to provide such detailed information on the construction dates of these sites. Until then, in fact, archaeological dating worked by approximations and resemblances with other localities for which, by some lucky accident, the exact date of construction was known. Findings in hand, Douglass demonstrates that dendrochronology can make a fundamental contribution to archaeology. The system works and it is simple. The instrumentation needed to extract the sequence of growth rings is, in fact, limited to the Pressler borer, a small drill that extracts cylindrical samples just a few millimeters in diameter, without any need to damage the tree or the beams being studied. After a modicum of training, anybody can use the borer to take excellent samples.

In 1928, working on living trees and wooden artifacts found in the ruins of the Colorado Plateau, Douglass uses this system to formulate two chronologies, each of them more than five centuries long. The problem is that they are not contiguous. The one based on living trees begins in the present and goes backward for five centuries and, in this case, each ring is dated precisely. The one based on dead trees or artifacts, though perfectly sequential, is not dated. Douglass has no basis for tying this second chronology to real events. There is, obviously, a time gap between the two chronologies that impedes him from obtaining a single, very long chronology, but there is no way of knowing if the time span of the gap is a few years, some decades, or even several millennia. The search for one or more wood specimens able to fill the gap becomes crucial. What is certain is that the possible applications of a sequence of this kind would be countless. For example, the little-known ruins of the ancient civilizations dotting the Colorado Plateau could be dated. This would be the first real and important application of Douglass's discovery. Something completely different from his initial research on the cyclicality of solar activity, but very interesting all the same. Maybe even more interesting.

The approach used by Douglass to fill the gap between the two chronologies is two-pronged: first, continue his expeditions in search of wood specimens taken from the ruins of the ancient settlements in the Southwest, and second, find a tree old enough to let him extend backward his chronology based on living trees. Trees that old do exist, not on the Colorado Plateau but rather in the nearby High Sierra, home to the giant sequoia

(*Sequoiadendron giganteum*), some of which have been around for more than three thousand years. If it were possible to find an overlap between the chronologies of the high plateau and those of the High Sierra, the problem would be solved. Unfortunately, the climatic conditions of the High Sierra are so different with regard to precipitation and temperature that the two series cannot, in any way, overlap. The only chance remaining is to find some wooden remnant among the ruins.

So in 1929, Douglass sets out on an expedition whose only purpose is to find a specimen that would span the gap between the two chronologies. He has no confidence that this attempt will lead to any useful results. His chronology is based on living trees and goes back to 1270, while the other chronology, obtained from the wooden artifacts, covers a period of another five centuries. The archaeologists keep telling him that the ruins on the high plateau are from a very ancient epoch. According to their estimates, the interval between the two chronologies is at least five centuries. The chances of finding specimens capable of spanning this interval appear to be very limited. But, on June 22, 1929, a beam labeled with the serial number HH-39 extracted from the site at Show Low, Arizona, succeeds in uniting the two chronologies and demonstrates that the archaeologists' predictions were wrong. The sites in the Southwest are much more recent than had been expected: The interval spanned by specimen HH-39 is actually less than one hundred years.

After decades of work, Douglass now has a continuous chronological series going back to 700 CE. This allows him to date, with total certainty, all the ruins present in the American Southwest: Mesa Verde and Canyon de Chelly date back to the thirteenth century; the Aztec ruins were built during a period from 1111 to 1120. Pueblo Bonito was built toward the end of the eleventh century.[4] In 1929, immediately after formulating the entire chronological series, Douglass notes that "the two chronologies were definitely matched and their unity confirmed by Beam HH-39, which in American archaeology is destined to hold a place comparable to Egypt's Rosetta Stone."[5]

A new method had been discovered that was practical, absurdly simple, and, in principle, applicable in the dating of any ancient human civilization, provided one has the correct sequence of growth rings in trees. You might expect that this discovery would have provoked an instant request of analogous services to date the appearance of ancient civilizations everywhere on the planet. Yet, after the triumph of dating in the American Southwest, it seems that there are no other places on earth where the same system can be used. Nothing more happens.

The amazing aspect of this whole adventure is that, even as late as 1936, Douglass is convinced he can use dendrochronology to obtain data on the cyclicality of solar activity, and he continues, undaunted, analyzing tree rings in the certainty that the data retrieved can be used to extract important information about solar cycles.[6] For Douglass, the question of archaeological dating is secondary compared to the possibility of figuring out the solar cycle.

Nevertheless, he is well aware of the importance that his research could have in disciplines far afield from astronomy. So in 1937, after founding the world's first laboratory devoted to the study of tree rings (the Laboratory of Tree-Ring Research of the University of Arizona at Tucson), he becomes the first person to teach dendrochronology, a discipline of which he must be considered, for all intents and purposes, the founder. In those years, his dating of the chronological series from the Colorado Plateau is extended, by way of more specimens, all the way back to the first century BCE. Then, there seems to be no way to extend them any further, and the studies come to a halt. In the Southwest, there are no ruins older than the ones already found, nor can the series of the High Sierra sequoias go beyond the three thousand years reachable with the oldest living examples. For the next twenty years, nothing of any importance seems to be happening in the field of dendrochronology.

But this state of affairs couldn't last. And, in fact, it didn't. The Tree-Ring Research Laboratory soon welcomed Edmund Schulman, a young assistant who would shortly become Douglass's primary collaborator and successor. The two researchers were very different. Douglass was the representative of heroic and intuitive science; Schulman was more rigorous and analytical. But they had something fundamental in common: the steadfast conviction that dendrochronology could provide vast amounts of information to science, and a great love for exploration and field work.

Starting in 1939, and for the rest of his life, Schulman spent every summer searching for the oldest living trees in the Southwest. The great majority of his colleagues were convinced

that it was impossible to find a living tree older than the so-called
BC barrier, or before the year zero of the contemporary era.
Nevertheless, Schulman stubbornly carried on his research. For
many years, he was unable to find a tree more than 1,700 years
old. Then, in 1953, returning from one of his camps in Sun Valley,
Idaho, where a year earlier he had found a 1,650-year-old *Pinus
flexilus*, he decided to make a detour toward the White Mountains
of Nevada and California, where a lot of people had told him there
were some unusually old trees.

Schulman didn't have much confidence in these rumors.
For decades he had been checking out trees he had been told
were extremely ancient but that, in the end, turned out to be
much younger than expected. This time was different. The area
that had been indicated was not completely unknown to him.
It was a forest of *Pinus longaeva*. Some years before, a ranger by
the name of Alvin Noren had identified an enormous specimen
there called "the Patriarch," about whose age Schulman had
fantasized a lot before finding out that it was only 1,500 years
old. Schulman was fascinated by the forest of *Pinus longaeva* and
sensed its enormous potential. For all of his remaining years,
that part of the White Mountains would be the place where he
concentrated all of his research.

Schulman spent the summer of 1957 in the driest part of the
forest, which will later come to be known as "the Methuselah
Walk." The conditions there are truly extreme, with rocky
outcroppings and minimal precipitation, perfect for his search
for a tree born before the birth of Christ. His findings went far

beyond his wildest dreams when he identified Pine Alpha, the first tree known to be more than 4,000 years old.[7]

The area explored by Schulman couldn't have been better for his purposes. From 8,000 to 12,000 feet above sea level, with limited resources and an adverse climate, pine trees grow to the upper limit of their range. Ideal conditions for trees to become champions of longevity. The soil, originally dolomitic limestone, is so deprived of water and nutrients that the trees grow stunted and contorted but live thousands of years. Giant bonsai is the first description that comes to mind to anyone who has seen them up close. Schulman speaks of "longevity through adversity"[8]; the same species, in the more favorable conditions found in other, even nearby regions of California, don't live anywhere near as long as the record-setting trees of the White Mountains.

It is by now an established fact, even with regard to animals, that conditions of caloric restriction significantly increase individual longevity. Life seems to prefer limitation over abundance and this area identified by Schulman appears to be the shining example. His discoveries come one after another. While in his laboratory studying a specimen extracted from one of the pine trees, he realizes that the tree it came from, which he calls "Methuselah," is more than 4,600 years old. It is the oldest known living being.[9] A few years later in the same area, other researchers will discover a 4,900-year-old pine tree called "Prometheus," whose fate we shall consider soon.

Thanks to these champions of longevity, dendrochronological research breathes new life, with expectations very different from those it had just a few years earlier. The dry, cold climate of the area had also kept dead trees from rotting, conserving them for millennia. Between living trees and dead trunks on the ground, an uninterrupted chronological series of 9,000 years is no longer a pipe dream. In a very short period, Schulman discovers twenty other trees more than 4,000 years old, but he doesn't have time to study their chronologies because—as fate would have it—he dies prematurely at the age of forty-nine, having lived a life a hundred times shorter than that of the creatures he loved and studied.

In 1964, Donald Currey, a rookie researcher fresh out of the University of North Carolina, adds his name by right to the history of dendrochronology, but for reasons having nothing to do with merit. In that same year, Currey is awarded a scholarship from the National Science Foundation to analyze, with the help of dendrochronology, the so-called Little Ice Age—a cold epoch in the history of the planet, conventionally restricted to the period between the sixteenth and nineteenth centuries, although some experts prefer a more extended period going from 1300 to 1850. We will see that this specification regarding the beginning and ending dates of this period is important.

The activities for which Currey received funding included a summer expedition among the *Pinus longaeva* for the gathering of data useful to his research. To facilitate the performance of this task, Currey requests and receives authorization from the Forest Service to cut down a tree so he can study the chronology of its rings. Why it was necessary to cut down a tree remains a mystery. By then, the Pressler borer had been a very well-known instrument for decades and part of the standard equipment for any expedition to gather dendrochronological data.

In any event, in the summer of 1964, Curry decides to visit a population of *Pinus longaeva* on Wheeler Peak, in eastern Nevada, which despite being located outside of the basin of the White Mountains had already been widely studied by researchers. Apparently ignorant of the findings of the preceding missions, Currey starts gathering specimens from the trees on site, and notices that many of them are more than 3,000 years old. He is particularly attracted by one tree, which he will call WPN-114. He uses a borer to take four different samples, but nevertheless he can't manage to get a continuous sequence. He breaks two drills in the attempt to obtain additional samples and then, unable to get what he needs, he asks the Forest Service to authorize him to cut down the tree. The Forest Service approves the request and, as if that weren't enough, offers its own personnel and equipment to do the job better.

On August 6, 1964, WPN-114 is cut down and its quartered sections are analyzed to determine its age. The findings are shocking and disconcerting. The tree turned out to be not only an old specimen but the *oldest* of all the living beings on the earth. Currey, with the help of the Forest Service, in an unprecedented combination of negligence, incompetence, and stupidity, had managed to kill Prometheus, the deacon of life on the planet.

How was it that the Forest Service and the rookie Currey didn't know about Prometheus? Why did the Forest Service authorize the cutting down of a tree without there being any real scientific reason to do so? And, above all, why was Currey, whose task was supposedly to analyze data regarding the Little Ice Age, interested in trees more than 4,000 years old? None of these questions has ever been answered. News of the cutting down of Prometheus was kept secret for years and, when it finally began to circulate, responsibility for what happened was difficult to ascertain. What we do know is that no one paid for this crime against life. Even worse, nobody even believed that what happened was a crime. After all, they'd only cut down some *timber*.

At this point in the story of astronomers, age-old trees, forests, and tree-ring dating, a towering figure comes on the scene: Willard Frank Libby. The winner of the Nobel Prize in Chemistry in 1960, at first glance Libby is hard to fit in with the rest of the cast. On a closer look, however, he has a few things in common with our other characters, starting with being born to a family of farmers from Colorado, the state that has been the stage for the entire story we have been telling up to now.

From the outset of his career, Libby was interested in the study of radioactivity, both natural and artificial. He built some highly sensitive Geiger counters and, inevitably, when the United States entered World War II in 1941, he became part of the Manhattan Project, the group of scientists that developed the atomic bomb. At war's end, he returned to his studies of radioactivity and accepted a professorship at the newly created Institute for Nuclear Studies at the University of Chicago. Meanwhile, in 1939, Baron Serge Korff had discovered something that will become fundamental both for Libby's career and for the rest of our story: that cosmic rays, upon arriving in our atmosphere, generate neutrons which, when they interact with nitrogen in the air, produce carbon 14 (^{14}C).

What makes carbon 14 interesting is that this carbon isotope is not stable but decays (transforms into something else), with a half-life of 5,730 years (over 5,730 years, the amount of carbon 14 is reduced by half). Libby understood that every living being, whether plant or animal, absorbs carbon 14 during its life and that, from the moment it dies, this carbon 14 begins to decay, being cut in half every 5,730 years.

Now, if some of you who are unaccustomed to scientific explanations are starting to feel their hearts flutter, please bear with me. There is absolutely nothing technical about what I am about to tell you. Let's get back to our story.

The carbon 14 in the atmosphere is mostly in the form of carbon dioxide and, as such, is absorbed by plants through the process of photosynthesis and transformed into organic matter. So every plant (and vegetable material) has a certain amount of carbon 14. The same is true for animals, whose carbon comes from the food chain and, therefore, in the last analysis, from plants, which are its base. What I mean to say is that whatever an animal eats—plants or other animals—the carbon that it is made of always comes from plants. Okay, the worst is over.

So, as Libby realizes, every organic material must possess some carbon 14 in lower and lower quantities depending on the time that has passed from the death of the organism that produced it. A piece of wood, a cloth, leftover food, paper, animal bones, a human skeleton, anything with organic origins is datable by measuring the amount of its residual carbon 14.

The first problem that Libby has to solve is proving that carbon 14 is actually present in organic material. This is not at all an easy problem to solve. In fact, the amount of carbon 14 can only be measured by the radiation it produces, but this is of such low intensity that there is no system sufficiently sensitive to detect it. Libby, however, as mentioned earlier, has an extraordinary training in the construction of Geiger counters, and, without getting into the details, he gets to the bottom of the problem and succeeds in demonstrating the presence of carbon 14 in living matter.[10] Having found the method for measuring the extremely weak radioactivity produced by the decay of carbon 14, Libby has put his hands on a powerful system for dating any organic substance.

The next step is to test the validity of the new technique by analyzing specimens whose age is already known by other means. To do this, he uses two specimens taken from the tombs of two Egyptian kings, Djoser and Sneferu, dated to 2525 BCE, with a margin of error of plus or minus 75 years. Libby's carbon 14 measurement dates them to 2800 BCE with a margin of plus or minus 250 years.[11] This was an excellent result. In 1952, with the publication of his *Radiocarbon Dating*,[12] the technique is now ready to be used to revolutionize entire disciplines like archaeology or paleontology, and any other field devoted to the study of ancient artifacts. By 1960, more than twenty radiocarbon-dating laboratories have been set up all over the world. As the Nobel Prize committee wrote on awarding Libby the chemistry prize in that same year, "seldom has a single discovery in chemistry had such an impact on the thinking in so many fields of human endeavor."[13]

Compared to dendrochronology, carbon 14 dating has a lot of advantages: It can be applied to any organic material, not only to wood; it does not rely on local chronologies; it requires much less material to perform its dating procedure; and so on. Nevertheless, when applicable, dendrochronology is much safer and less subject to error caused by outside factors, as happens with carbon 14 dating, as many experts would soon come to realize. In fact, the carbon 14 dating technique was not without its weak points. The first measurements performed in the 1950s yielded results that were so surprising and different from what could have been anticipated that many archaeologists believed, without a doubt, that the datings were irremediably wrong.[14]

It must be recalled that, in the first half of the last century, most archaeologists were firmly convinced that the great megaliths, like Stonehenge, for example, or other megalithic complexes in Spain and other European countries, were subsequent to the structures of Mycenaean Greece (1600–1100 BCE). However, according to the measurements carried out with the carbon 14 method, these monuments were given earlier dates, and often much earlier, than what was expected. Stuart Piggott, a famous British archaeologist, claimed in his book *Neolithic Cultures of the British Isles*, published in 1954, that the beginning of the Neolithic age in the British Isles was not to be considered antecedent to 2000 BCE. Piggot saw his claim disputed by radiocarbon datings, which pushed this date back by at least a thousand years. He did not take this at all well, and declared that the newly obtained datings, "being archaeologically unacceptable," indicate that the radiocarbon system is subject to error.

In a sense, he was right, but what he certainly could not imagine was that those errors were not, in fact, in the direction he thought. The Libby method is based on the assumption that the amount of carbon 14 in the atmosphere remains constant over time. This assumption is totally arbitrary, without any corroborating evidence, and does not have a large following in the scientific community. There are many skeptics who do not think this is the case at all. On the contrary, they are convinced that, given the variability with which cosmic rays strike the atmosphere, a constant quantity of carbon 14 is highly unlikely. Luckily, plants, yet again, come to our aid. Thanks to good old reliable dendrochronology, plants are able to provide a chronological series of more than 9,000 years against which the radiocarbon technique can be calibrated. The Laboratory of Tree-Ring Research offered Libby all of its experience and, above all, the uninterrupted chronological series from the *Pinus longaeva* of the White Mountains. He could use the chronology as a model with which to measure eventual deviations from the norm in the datings obtained with carbon 14. This was an elegant and solid system. Nevertheless, in the beginning, Libby did not like it at all. Why? We will never know for certain, but I would bet that it must have seemed to him a less than truly scientific approach. However, the criticisms of his early datings keep getting stronger and the scientific community is more and more convinced that assuming a stable level of carbon 14 in the atmosphere had been a real gamble.

Libby needed to obtain confirmation of his technique and there was no other way to obtain it except by way of dendrochronology, so he relented and gave his consent so the 9,000-year series of growth rings from the *Pinus longaeva* could also be measured with the carbon 14 technique. The findings confirmed that the amount of carbon 14 in the atmosphere varies over time and, consequently, that the findings obtained with radiocarbon dating had to be corrected. By a lot.

The curve obtained from the dendrochronological data shows two types of variation from a straight line: a fluctuation with a period of about 9,000 years and a much shorter variation with a period of decades.[15] The first variation is very probably caused by the change in the intensity of the earth's magnetic field.[16] The shorter variations, about which we will have more to say in a moment, are known today as the De Vries effect, from the name of its discoverer, Hessel de Vries. A renowned Danish physicist, De Vries, if he hadn't committed suicide in 1959, after killing an analyst with whom he had a relationship, would have stood a good chance of sharing the Nobel Prize with Libby in 1960.

In subsequent years, many other sources of atmospheric carbon 14 variation were discovered. Some of them derive from recent human activity such as the combustion of fossil fuels (Suess effect)[17] or atomic explosions, but their influence on the dating of ancient specimens is nil. Conversely, the large variations of atmospheric carbon 14, found thanks to dendrochronology, are extremely important. Starting with the first millennium BCE, the calibration curve tilts rather rapidly lower due to higher concentrations of carbon 14 in the atmosphere. The result is that a date calculated with carbon 14 to be around 2400 BCE but not calibrated, has to be corrected, pushing it back 600 years earlier, while a date calculated to be around 3100 BCE has to be corrected by moving it back all the way to 1,100 years earlier. Poor Professor Piggott was right in pointing out that something was wrong with Libby's first carbon 14 datings. What he couldn't imagine was that the correction would move the date even farther back in time. Once corrected with respect to Libby's initial estimates, the dating of British Neolithic artifacts had to be predated by another one thousand years. A true revolution for European archaeology.

Today, carbon 14 dating has no more secrets. All that's left to be discovered are the causes of those small variations, which have a period on the order of decades and are known as the De Vries effect. One of the more likely explanations is that they are caused, oddly enough, by variations in solar activity. More solar activity means more radiation reaching the earth, and so more carbon 14 in the atmosphere.

Douglass, the father of dendrochronology, can rest in peace. It wasn't climate that was influenced by solar activity but the quantity of atmospheric carbon 14. But whether it was one or the other, he was still right: In either case, traces of solar activity were to be found in tree rings.

I think of the Garden of Eden as a tropical forest . . . in which case the tree of knowledge would surely be a banana tree.

Six

PLANTING
KNOWLEDGE

*Two portly sows are trotting up behind this carriage, and a select
party of half-a-dozen gentlemen hogs have just now turned the
corner. Here is a solitary swine lounging homeward by himself.
He has only one ear; having parted with the other to vagrant-
dogs in the course of his city rambles. But he gets on very well
without it; and leads a roving, gentlemanly, vagabond kind of
life, somewhat answering to that of our club-men at home. He
leaves his lodgings every morning at a certain hour, throws
himself upon the town, gets through his day in some manner
quite satisfactory to himself.*

The author of this account is Charles Dickens, on a visit to
New York in 1842. Walking along the streets of the city,
the thing that strikes him most is that these pigs

*are never attended upon, or fed, or driven, or caught, but are
thrown upon their own resources in early life, and become
preternaturally knowing in consequence. Every pig knows where
he lives, much better than anybody could tell him. At this hour, just
as evening is closing in, you will see them roaming towards bed
by scores, eating their way to the last. Occasionally, some youth
among them who has over-eaten himself, or has been worried by
dogs, trots shrinkingly homeward, like a prodigal son: but this is a
rare case: perfect self-possession and self-reliance, and immovable
composure, being their foremost attributes . . .*[1]

This is not the literary invention of a great writer. As unlikely
or bizarre as it might appear to our eyes, the scene described is
absolutely true and numbers among its causal factors . . . bananas.
I'm talking about the fruit, not the tree that produces it, and not
even the whole fruit but only its humblest, most external part: the
skin. The part we throw away.

You can't imagine how many enchanting stories can be told
about this modest waste product. Since the time I began taking
an interest in banana skins, I have discovered literally dozens.
Starting with the story that explains the reason why "slip on
a banana skin" has become in so many languages around the
world an idiomatic expression for an embarrassing failure. Why
a banana skin and not the skin of some other fruit? Orange
peels, watermelon or cantaloupe rinds, or peach skins, to name
the first that come to mind, don't seem, at least on first glance,
any less slippery than a banana skin. Yet everywhere you go, the

talk is only about banana skins. The banana skin is the only fruit covering to be directly cited in an idiomatic expression. True enough, skins are used in a lot of phrases: "lose your skin," "save your skin," "slip me some skin," "get under my skin," "have a thick skin" are just some of them. But the only fruit skin to have an idiomatic expression all its own is the banana skin.

This apparently unimportant question began to haunt me in the spring of 2014, during one of my frequent visits to the laboratory in Kitakyushu, on the southern island of Japan. On these occasions, my scant knowledge of the Japanese language— no, let's just go ahead and say my absolute ignorance of it— combined with my students' natural indisposition to speak with someone older and higher up in the hierarchy than them, limits my social life to very few interactions, mostly of a scientific nature with colleagues. This really bugs me. I have a great admiration for the Japanese culture, and trying to understand the reasons for its charm without being able to speak with those directly involved is, as you can imagine, a decisive limitation. In any case, there is not much I can do about it. Even if I learned Japanese perfectly, I doubt I would understand much more about the culture. As a foreigner, the contents and concepts that I have access to are few and superficial.

This would all be unbearable for me if it weren't for my friend and colleague Tomonori, who without his even knowing it remains, thanks to his friendliness and our countless discussions, my main point of access to this mysterious and distant civilization. Tomonori, Tom for me, is a totally anomalous Japanese; we might call him a mutant. For example, he is the only Japanese I know who is able to say, without seeming awkward, "Well, look what time it is! Let's go have a drink." In the mouth of any other Japanese, this phrase would sound off-key or out of place or, in the best of cases, oafish. Although many of them give it a try, the effect is always the opposite of the warm openness of an invitation; it's as though they're saying it against their will. It's a little like when we Westerners try to imitate their bows. We just can't do it. We make them laugh. As Tom says, "Forget the bow, it's not your thing." Tom, as I was saying, is a mutation. For him it's natural to go out together for a drink. What a lucky coincidence it has been for me to become his friend! I mean, how many mutant Japanese can there be? I have never met any others except for Tom. As far as I know, he may be the only one. And it is thanks to our chats that I have had a chance, over the years, to understand something about the Japanese people.

One evening, Tom called me to invite me out for a drink. Essentially, going out for a drink means going, at the end of the workday, from the laboratory to his favorite tavern—I wouldn't know how else to define that dark little dive where we go to drink a beer and eat micro-portions of various Japanese delicacies. The interesting thing about these evenings out is that usually, after our first beers, some topic of discussion pops up about things, concepts, commonplaces that are obvious for us Westerners but for the Japanese make no sense at all. Or vice versa. That night, the topic was banana skins. I was saying that somebody, I think a politician, had slipped on a banana skin when Tom, looking up from his beer, gave me a big smile and asked, "And did he get hurt?"

"Get hurt?" I replied. "What do you mean? He didn't really fall down, it's a metaphor."

That's how I came to find out that for Tom—I won't risk saying for all Japanese—"slipping on a banana skin" had only a literal meaning and was without any metaphorical implication whatsoever. This left me dumbfounded and gave rise to a long and amusing discussion about why this manner of speaking was so easily grasped by the rest of the world's population. In these chats at the bar, the unwritten rule for Tom and me is that each of us has to take aim at what seems to him to be the weakest link in his adversary's logic. It goes on like this on a rising tide of beer for several nights, until one of us comes up with some crushing evidence in support of his position.

Right from the get-go, Tom's point of attack was this: "Why a banana skin and not the skin of some other fruit? Is a banana skin that much more slippery than all the others?" I forgot to mention this, but Tom is an extraordinary chemist and so he tends to base all of his bar talk on the need to have incontrovertible data regarding the nature of materials. To give you the idea, with regard to this story about banana skins, he would never have accepted the reasonableness of the idiomatic expression unless and until I demonstrated the greater slipperiness of the banana skin.

The incisiveness of Tom's observations immediately set me back on my heels. Where the devil could I ever hope to find solid data on the slipperiness of banana skins compared to other fruits? In the absence of such data, whatever evidence I might offer in support of my position would only make it easier for Tom to gloat. Tom was enjoying this situation more than any other beer debate we'd ever had. Believing he had laid his hands on an invulnerable argument, he let me go on talking for hours about bananas—about their nutritional and social value, about the alternative uses that might be invented for their skins, about the functioning of parthenocarpy[2]—but no matter what information I came up with, interesting as it might have been, he would not budge one inch on the question of why banana skins were to be considered the ne plus ultra of slipperiness.

After each of my ramblings, Tom would look at me with an innocent air and say, "And so?" The thing drove me up the wall so much that just to wipe that smile off his face I started reading whatever I could find about banana skins in the hope of finding something to hang my hat on to move the discussion my way. If nothing else, the research would have been worth it for that, and what follows are some of the things I learned.

First, however, despite my initial statement that I would only talk about the banana skin, I have to give you a quick run through the basics of the history of the banana tree and of its fruit, which the extraordinary skin comes from. The name banana has been given to numerous species that belong to the genus *Musa*.[3] Of the seventy or so species of which the genus is formed, almost all of them are able to produce edible fruit. Nevertheless, nearly all seedless bananas (partenocarpic) that we consume come from just two wild species: *Musa acuminate* and *Musa balbisiana*, both first described in 1820 by a lawyer from Turin, Luigi Colla.[4] The banana is not a tree, as it would appear to be, but rather a mammoth herbaceous plant able to produce berries (the proper category for the banana fruit), which are among the first fruits that were cultivated and harvested by humans. Indeed, in many Asian traditions it is the banana, and not the apple or the fig, that is the true tree of knowledge. Whether or not it is the tree of knowledge, the banana is certainly one of the plants with the longest relationship with humans.

The first traces of its cultivation go back to Southeast Asia and to Papua New Guinea, where recent findings of archaeological and paleoenvironmental studies seem to suggest that the cultivation of bananas dates back to 8000 BCE.[5] As with many plants that are important for the human diet, this must not be considered the unique origin of its cultivation. In all probability, other species were domesticated independently in other regions of Southeast Asia.

Okay, now that we know at least the basics, let's see if we can understand something about why banana skins became famous. Their history is strictly tied to the history of commerce in North America in the second half of the nineteenth century. At that time, in a city like New York, few people could afford bananas. Their cost was so high that it was comparable to the cost of caviar. Bananas were thus a sort of status symbol to be put on display at evening galas or to impress your friends. The reasons for this are not hard to understand. In the mid-nineteenth century, transporting bananas from where they were cultivated all the way to New York, without their arriving completely transformed into a putrid mush, must not have been easy. The transcontinental railroad was still a dream, and the only means of transport efficient enough to ensure a minimum chance of success were sailing ships, which, as fast as they were, took three weeks to make the voyage from Jamaica to New York. Not too long a time, but still not short enough to guarantee a substantial presence of bananas at the right stage of ripening to satisfy the refined New York market.

Then everything changed. The banana market turned upside down, owing in part to steamships, which were rapidly replacing the old schooners, and in part to the Boston Fruit Company's clever idea to harvest bunches of green bananas and ship them in special refrigerated holds, first with ice and then with increasingly sophisticated systems. The ripening process was then finished off in the fruit and vegetable markets of the bananas' final destination. In a short time, the number of bananas able to reach the North American market in perfect condition multiplied, with a proportionate reduction in their price. For centuries, bananas had been a delicacy reserved for the wealthy. Over a very short span of years they were transformed into the opposite: street food. This is the period that launched the great commercial success of the banana. Today, it is the best-selling fruit in the United States and the fourth-ranking staple in the world, behind rice, wheat, and corn.

In New York at the end of the nineteenth century, bananas were sold on the corners of the main streets and their price was so low that they became a popular food item. The Boston Fruit Company (later to become United Fruit Company and, later still, Chiquita) inundated the cities of North America with excellent bananas, and everybody involved—industrialists, traders, and consumers—is happy. As long as we exclude the producers of bananas, whose shameful exploitation has never stopped since bananas arrived in North America transported by the Spaniards, for everyone else the circulation of bananas turns out to be a great deal.

Although the growing consumption of bananas improves, by a lot, the diet of a population used to much less nutritious low-cost food, it also raises the question of how to dispose of the growing quantity of waste that this consumption produces. In less than a generation, banana skins became one of the most common pieces of refuse scattered over the streets of New York. Not that the problem can be attributed only to bananas, obviously. Late nineteenth-century New York by no means excels in the cleanliness and orderliness of its streets. On the contrary, garbage disposal is reduced, for the most part, to its simple abandonment on the street. There is no agency for waste management, no system of garbage collection. Refuse is blithely left out on the street in such quantities that it forms insuperable traffic barriers. Newspaper stories tell of continual traffic detours owing simply to the necessity of avoiding streets made impassable by the sheer quantity of garbage. Entire neighborhoods are considered off-limits due to their poor sanitation. Five Points in Manhattan, for example, for decades a prime example of a

Western slum, is considered a no-go zone, not only for the crime rate typical of inner city slums but also for its disastrous sanitary conditions, which provokes continual epidemics and a sky-high infant mortality rate.[6]

Anyway, even leaving aside the slums, the whole city is besieged by garbage. Somehow, the refuse problem has to be solved. What to do? One of the remedies conjured up by New York's municipal authorities demonstrates, in its simplicity, all of the practical genius of Americans. How is garbage handled on farms? Simple: You use it to feed the pigs. So why not do the same thing in the city? No sooner said than done. Tens of thousands of pigs are transported from the country into the city, entrusted to custodial families, left free to circulate as they please on the streets of New York, feeding on the city's refuse. Today this would seem like a desperate solution, yet think of the enormous practical advantages: the removal of most of the city's refuse and its conversion into prized pork.

If it hadn't been for certain aspects of, let's say, decorum, about which I'll say more in a moment, the solution yielded excellent results. Not only did the garbage on the streets of New York drastically diminish but at the same time thousands of indigent families now had at their disposition a source of highly nutritious food. Just think if today, in an age devoted to recycling and the circular economy, someone were to invent a machine able to do even a fraction of what a pig can do. Sure, the downside was that there were thousands of pigs on the loose, as we have already seen in the account of Charles Dickens. Those who couldn't stand them called them "walking sewers," but for everyone else they were salvation.

Obviously the situation couldn't last long. Even despite the pigs' efficiency in cleaning up the city, problems soon start to arise. They are accused of attacking children, of defecating on people, and of spreading terrible diseases like cholera. Serious problems, certainly, but not irresolvable. The drop that tips the bucket is something else. New York men can't put up with pigs coupling freely on the streets. They worry that this lascivious aspect of their nature might induce their irreprehensible wives and daughters to think with more indulgence of the pleasures of the flesh. This risk, obviously, is intolerable. The fate of the pigs is sealed and, after just a few years, they are banned. New York is freed of its pigs, but it is once again besieged by garbage. And by banana skins.

The quantity of banana skins on the streets is so great that the number of accidents caused by falls, skids, slips, and slides becomes a proper emergency. Newspapers of the time are full of accounts of ruinous falls with consequent injuries ranging from sprains and fractures to, in the most unfortunate cases, even death. This is the period of the first slapstick comedy gags in vaudeville and then in the early silent movies. Ruinous slips on banana skins are first turned into sublime art by Charlie Chaplin, who introduces the gag to cinema, and later by Buster Keaton and dozens of others.

Banana skins are good for a laugh but they can also hurt you. The number of accidents is so high as to demand drastic measures. On February 9, 1896, Theodore Roosevelt, then the chief of the New York police, attempts to put a stop to the huge number of accidents afflicting his fellow citizens on a daily basis. He issues an ordinance prohibiting banana eaters from throwing the skins on the street and requiring sellers to print and display the ordinance prominently in their shops. Offenders are subject to a fine of up to ten dollars and in serious cases even imprisonment. That's not all. To implement the ordinance and enforce it effectively, Roosevelt gives full powers over the cleanliness of the streets to the former Civil War colonel George Waring. An officer from the old school, who had already won acclaim for his modernization of the sewer system in Memphis, ending an epoch of continual epidemics, Waring seemed like just the right person to clean up the garbage in New York, banana skins and all. The colonel wastes no time and in short order transforms the city's street cleaners into a sort of militia, outfitted with their own special immaculate uniforms and instilled with iron military discipline. The operation achieves the desired result and New York is finally liberated from pigs, garbage, foul odors, and banana skins.

Although New York had solved its problem with banana skins, garbage, and all the rest, there was still the open question of why it was the banana skin, and not the skins of other fruit, that became the object of thousands of gags, to the point of becoming synonymous with the very idea of slipperiness. In short, the question posed to me by Tom—Why the banana skin and not the orange peel or the watermelon rind?—was still valid.

I was getting ready to leave Japan. Despite weeks of discussions and data regarding the history of bananas and their skins, I still hadn't made even one step forward in the explanation of their slipperiness. Tom continued to listen to me attentively over our beers while I told him about the development of sanitation practices in New York, only to look at me in the end with a little smirk and mumble his "And so?" It was turning into an obsession. I had learned a lot of things about banana skins but nothing about their slipperiness. About that, I knew exactly as much as I'd known before. The difficulty was that there was no data on the slipperiness of bananas and I knew that, until I was able to produce some numbers, Tom would keep needling me with his "And so?"

Not wanting to concede defeat, I decided I would measure the slipperiness of banana skins myself. At Kitakyushu, I had a laboratory outfitted with all kinds of equipment and instrumentation and lots of collaborators capable of figuring out the solutions to the most desperate problems related to the world of plants. When it comes right down to it, I said to myself, banana skins are a vegetable material. Finding out a little more chemico-physical data concerning their composition certainly won't hurt us. So one day I called everyone in the laboratory to a meeting to discuss the problem. "Can anyone here come up with a test to measure the slipperiness of a banana skin?" I was asked to repeat the question a couple of times; they thought maybe they'd misunderstood me or that I was joking. I reassured them of my mental health and the need to have measurable data on the slipperiness of banana skins.

One thing I adore about my Japanese collaborators is that you don't have to spend hours convincing them of the reasons why it is necessary to do something. In Florence, for example, if I ask one of my researchers, "Could you please measure this or that for me?" I always have to be ready to explain the reasons for my request, the scientific motivations, or why I chose that specific measure instead of some possible alternative. It can get very tiresome, believe me. Okay, in part it's my own fault, I shouldn't complain. When they start working for me, I'm the one who always asks them the reason for things and tells them never to take anything for granted. Sometimes, however, I wish I had never done it; it seems like I've created a laboratory staffed with sophists. In any event, if I had requested the test on banana skins in Florence, I would never have heard the end of it.

Fortunately, I was in Japan. The youngsters reassured me: Within a week, we would have all the data I wanted. Immediately, the laboratory was seething with activity. One group studied the existing literature on banana skins, another the literature on measuring slipperiness, a third group went to visit a laboratory in the nearby school of engineering where an engine research center studied lubricants. In our laboratory, secret meetings were held with experts on various matters, and new instruments appeared on the scene along with bananas of various shapes and sizes and degrees of ripeness. In short, the customary Japanese efficiency was at work. The industriousness of my collaborators put me in a good mood. I was certain that in no time at all I would rid Tom's mug of that supercilious smirk.

Wrong! The days went by one after another without even the shadow of a finding. After their initial exultation, with each passing day the faces of my collaborators lost a little more vim and verve until they finally threw in the towel, declaring that they had failed to produce any significant findings. Now, I know what a lot of you are thinking, especially you engineers and physicists (I purposely mention you together to make you angry): What could be so hard about measuring the slipperiness of a banana skin? I don't want to bore you with the details, I'll just say this: You try it, and then let me know how it goes.

In short order, the laboratory returned to its normal activities, leaving me with my ignominious, solitary obsession for the banana skin. I was really down in the dumps. Tom's smirk haunted me. I even started thinking that the young people in my laboratory were in cahoots with him to not give me the findings I yearned for. I was that far gone, and then a miracle happened. One of those portentous events that reconcile you with the divinity. Looking for distractions from my yellow fixation, I had started leafing through a recently arrived scientific journal. There was an article about that year's Ig Nobels.[7] The awards are always a lot of fun, and I was having a good time reading the motivations for the awards in the various disciplines until I arrived at the Ig Nobel in Physics, and the journal almost dropped out of my hands. The 2014 Ig Nobel in Physics had been awarded to a group of Japanese researchers, the adored Kiyoshi Mabuchi, Kensel Tanaka, Daichi Uchijima, and Rina Sakai of the University of Kitasato, *for having measured the quantity of friction between a shoe*

and a banana skin and between a banana skin and the floor, when a
person steps on a banana skin that's on the floor!

I read the article numerous times. I couldn't believe it. There
was something truly otherworldly about it. Some Japanese had
performed exact measurements of the slipperiness of banana
skins. How wonderful! How could I not love these people! The
article in question had been published in *Tribology Online*,[8] a
publication I had never even heard of and the meaning of whose
name I didn't understand. Can you believe that there exists an
entire discipline, tribology, which studies the slipperiness of
things? What a fantastic sensation. After weeks of tribulation,
tribology had given me a chance for vindication.

The article was exhaustive and provided all the information
I hadn't been able to find a trace of for weeks. Now I knew how
slippery a banana skin was. The friction coefficient of a normal
shoe sole on linoleum is 0.412, while the friction coefficient of a
banana skin was one-sixth as much, or 0.066. What's more, the
article provided data on the friction coefficients of apple peels
(0.12) and orange peels (0.22) and on other interesting pairs
of slippery materials. I discovered that the friction coefficient
of skis on snow is only slightly less than that of a banana skin
(around 0.04) and that the incredible slipperiness of ice on ice
had a friction coefficient of less than half of that of a banana skin
(right around 0.025). In brief, I now knew everything about the
slipperiness of bananas and just how much more slippery they are
compared to other fruits.

My first impulse was to charge into Tom's office, which was next to mine, and throw the freshly found article onto his desk. But I held back. Weeks of humiliating smirks had to find adequate satisfaction in a memorable final discussion. I stuck my head in the door of his office and nonchalantly asked if he had any plans that evening. He snapped at the bait. He eyeballed me with an unusual look that should have made me suspicious and, smiling, said, "Would you like to go out for a drink together? I love talking about banana skins." I assured him that I was delighted by the idea.

I was beside myself. I printed out a copy of the magnificent article published in *Tribology Online* and, way in advance of the agreed upon time, I went to the tavern to occupy our favorite table. Tom arrived right on time with his usual knapsack full of scientific articles. He set it on the floor beside him and casually took out a pack of publications in Japanese that he placed on the table as if they had no importance at all. We ordered our customary beers and started making small talk. I didn't want to be the one who raised the topic. I was waiting for him to bring up the story of the banana skins so I could finally crush him under the mass of tribological data.

I didn't have to wait long. Tom seemed anxious to resume the discussion, too. "So how is your research on the banana skins coming along? I hear you've had some trouble measuring their slipperiness." He must have heard about our laboratory attempts and the thing really amused him. Well! He wouldn't be amused for long. Without bothering to respond to his insinuations about the capabilities of my laboratory, I started showing off the data concerning the methodology, the friction coefficient of bananas, and everything I had learned about the fascinating universe of tribology, certain that it would make his smile disappear. But instead . . . the more I went on, the more Tom's smile broadened from one side of his face to the other in a seemingly irrepressible crescendo. Until I got to the point of citing the figure of 0.066 for the friction coefficient of banana skins, when he couldn't contain himself anymore, and broke out in an uproarious laugh. One of those never-ending laughs that leaves you breathless and that immediately become contagious. I couldn't resist either and, like all the other customers in the tavern, I started laughing with him without restraint. Never waste the chance to have a good laugh.

When we finally calmed down, and after completing the chain of toasts with our fellow beer drinkers that our group laugh had unleashed, Tom decided to tell me the reason for all that hilarity. As he pulled out of the pile of articles that he had deposited on the table the one on banana skins published in *Tribology Online* by Mabuchi et al., he confessed that he had known about their work all along. Indeed, his many interests related to the quality of materials certainly included tribology, of which he was a passionate admirer. In short, he confessed to me that he'd had the data in hand from the outset, but he hadn't wanted to concede defeat, and that he was sure I would never have read that one piece of research published by the unknown Japanese Society of Tribologists. Nobody could have foreseen that the 2014 Ig Nobel in Physics would be awarded for precisely that piece of research, making it famous. "Bad luck," he declared nonchalantly.

He took a sip of beer and he still hadn't stopped smiling at the story of his misdeed when, as he got up to go to the men's room, the low friction coefficient between his thin cotton socks and the wood floor provoked a clamorous and painful fall on his butt, in perfect Buster Keaton style. I managed to restrain myself from bursting out laughing and gave him a hand to get back on his feet. Suddenly he had turned serious. I asked him if it was because of the fall on his butt. "In part," he replied, "and in part because the fall is undoubtedly a sign from heaven." I looked at him to see if he was joking. He couldn't have been more serious. "What were the chances that you would come to find out about an obscure article published by a still more obscure Japanese scientific association in a scientific discipline that you didn't even know existed? Think about it. One in how many billion? Ten, one hundred?" I made an effort to console him, appealing to chance, coincidence, bad luck. But Tom didn't want to hear it. "No! I'm telling you, it's divine intervention. There is no other explanation. That's how you discovered that article and it was also divine intervention that made me slip and fall right here in front of you to punish me for my behavior." He added nothing else. His explanation satisfied me as well and for once we found ourselves in agreement about everything. Tribology giveth, I thought, and tribology taketh away.

The slipperiness challenge, besides bequeathing me knowledge of the friction coefficient between skin, sole, and linoleum, as well

as a myriad of other interesting factoids, which had enriched
my already enormous warehouse of little-known but useless
knowledge, had also made me the gift of a sort of Zen teaching.
If months of study are not enough to master even the most
banal information about something as marginal as a banana
skin, what chances could we ever have of comprehending the
world we live in? The fact that there was no chance at all filled
me with joy. Indeed, nothing worries me more than those
proclamations, which periodically, in sudden and irrepressible
bouts of presumption, bring someone to say that by now we know
everything about something. When I hear it said that we have
come to the end of art, or music, or physics, I start to feel my
stomach tightening from anxiety. The reason is twofold. On the
one hand, the enormity of human vanity never ceases to amaze
me and this, let's say, is the ethical side of it. But on the other
hand there is the more authentically egotistical aspect. When
we start saying that we know everything about something, my
first reaction is always one of loss. Just think if we really knew
everything about physics. All of a sudden, there would be no need
for physicists, we would never again read abstruse theories about
the multiverse, particles, black holes, or quantum entanglement.
Physics itself would vanish, becoming a simple postage stamp
conserved in the collection of knowledge.

Can you imagine how boring it would be to live in a world
where you knew everything? Well, don't despair. Luckily, that's
not likely to happen in the foreseeable future, and our knowledge
of banana skins is an obvious demonstration of that. Okay,

let's not talk about knowing their friction coefficient—that's something only for fanatics, I admit—but knowing how to peel a banana, that at least, you will agree, should be common knowledge. After all, we are talking about the fourth most consumed food item in the world! Yet this, too, seems to be completely unknown: We all attack the banana by breaking the petiole or stem, and then removing the skin. Wrong. The correct procedure, much more efficient and less wasteful in terms of energy, is the exact opposite. The banana is peeled starting from the opposite end from the stem. All you have to do is press the bottom end between thumb and index finger and the skin, as if by magic, splits into two convenient halves, which can then be removed. That's how chimps do it. Try watching some documentaries. They would never peel a banana the way we do. It's not efficient.

Now, if we don't even know how to peel a banana, how much can we possibly know about all the rest? Pretty close to zero, I would bet. So allow me to give you at least two other pieces of information about banana skins that strike me as decisive and that I hope will provoke a tingle of interest even in the most resistant of my readers. First: Bananas, skins included, are slightly radioactive. You weren't expecting that, were you? Actually, it is not such big news; all earthly objects that contain potassium—people, animals, and plants included—emit a minimum amount of radioactivity. To be precise, 31 becquerels per gram (Bq/gram), that is, 31 atoms of potassium per gram, decay every second.[9] The cause of this is that the isotopic mix of potassium contains 0.0117 percent of potassium 40, an unstable isotope with a half-life of around 1.2 billion years.

What makes the banana especially interesting from this point of view is its proverbially high amount of potassium, about half a gram (0.017 ounce) per banana. It was this high amount of potassium that, in 1995, suggested to some scientists the use of the banana equivalent dose (BED) to explain that radioactivity is part of our environment (obviously in very small doses) and that it mustn't always frighten us.

The BED is an informal, nonscientific measurement for which every equivalence, obviously, must be treated as necessarily approximate, but which is certainly useful when, for example, the public must be informed about radiation risk. Since the dose of absorbed radiation is measured in sieverts (Sv), a BED amounts to one ten-millionth Sv, or 0.0000001 Sv. The background radiation we are all subjected to amounts to about 0.35μSv per hour, or three and a half bananas. Proceeding along this path, we can say that an hour-long flight in an airplane at high altitude equals fifty bananas, sleeping next to someone half a banana, and an X-ray of an arm about ten bananas, while one hour of exposure to the worst conditions faced by the first responders after the Chernobyl disaster would amount to hundreds of millions of bananas.

But being blandly radioactive is not all. Our common banana skins possess another extraordinary characteristic: They are efflorescent if exposed to ultraviolet rays. This means that if we were to look at them through a camera lens able to see the ultraviolet band, the bananas would shine brightly. Not all of them, however, only the ripe ones. The efflorescence, in fact, is caused by a compound produced as a consequence of chlorophyll decay. In other words, a green banana does not shine, while a ripe banana sparkles like fireworks. For many animals capable of seeing the ultraviolet spectrum, this luminescence is a flash of good news. For them, coming across a shiny banana means having a clean shot at a ripe banana. This amounts to an extremely efficient advisory system for animals who, by eating the fruit, participate in the dispersion of its seeds.[10] In this way, the banana tree deters animals from feeding on the banana fruit whose seeds are not yet ready to be dispersed, directing their interest, instead, to the ripe ones, ready to be eaten, digested, and dispersed.

Obviously, it is important to recall, we are talking about uncultivated bananas, plants that still maintain intact their capacity to reproduce and disperse themselves. As we all know, the lack of seeds in the bananas we eat, strongly demanded by us consumers, prevents cultivated bananas from reproducing autonomously. For them, the only chance to reproduce is through the vegetative multiplication performed by humans. So if one day you should have at your disposal a UV-sensitive video camera and you see the bananas on your kitchen table shining bright, please know that nothing will come of it: By now they do not have any mature seeds to signal. You would be observing the vestiges of their happy past, when they were still free and worthy living beings, capable of evolving. In short, before we humans

STEFANO MANCUSO

transformed them into a vulgar means of production. The same
humans who fall for stories like the one I am about to tell you.

It seems unbelievable, yet for an unfortunately too brief
period of time, thanks to a well contrived and very funny
joke, our relationship with the plain old, sweet, yellow banana
changed into something new and unexpected. We are back in
1967, when, for a few months, a hoax—today we'd call it fake
news—convinced a lot of people around the world to smoke
dried banana skins in the certainty that they would stimulate
memorable hallucinogenic "trips" that were both legal and cheap.
We think that fake news stories are a specialty of our own time,
a consequence of the enormous power the Internet has given to
anyone who dreams up the wildest idiocy that our imaginations
are able to invent, always finding an audience disposed to
drink it down. But the truth is that fake news has always been
part of human communication. We could say that fake news is
connatural to humans. To be sure, today, thanks to the power of
the Net and the new technologies of communication, it takes no
time at all to make us believe that the earth is flat or that airplane
wakes are actually a system invented by the CIA to modify the
climate of the planet. But in 1967, the Internet did not exist, nor
did cell phones, social networks, or blogs, and even radio stations
and television channels were few and largely state-owned and -run.
In short, it was a time when news, still mainly disseminated by
newspapers, could not really travel very fast.

So then how can we explain that a multitude of people,
hip and not-so-hip, overeducated and undereducated, became

154

convinced in just a few days of something so clearly absurd as
the idea that bananas—the fruit most consumed by children—
contained potent hallucinogenic substances, and that it was
enough to smoke just a few skins to feel their psychedelic effects?
The event, which really does have something ominous about
it, could never have come about if a series of factors, partly
fortuitous, had not materialized simultaneously. But that is
exactly what happened.

First of all, let's quickly recall what was going on in 1967,
one of the mythical years of the 1960s counterculture. The
Vietnam War is in full swing, the Beatles have just released *Sgt.
Pepper's Lonely Hearts Club Band*,[11] the Doors and Pink Floyd put
out their respective debut albums, and Jimi Hendrix recorded
the incomparable *Are You Experienced?* In San Francisco, the
first issue of *Rolling Stone* is published and in Italy, Hugo Pratt
creates his graphic novel hero Corto Maltese. Movie theaters are
featuring Luis Buñuel's *Belle du Jour*, Arthur Penn's *Bonnie and
Clyde*, and Mike Nichols's *The Graduate*. Sexual liberation and
a wave of novelties in lifestyles, tastes, social relations, and the
individual and collective ethos are undermining the old order,
deluding many, for a brief and productive period, into thinking
that another world is possible.

A considerable part of this dream is associated with the use
of mind-altering substances. Everyone seems to be searching
for artificial paradises; all kinds of drugs or substances that
are thought to have psychoactive properties are tested both in
laboratories and, more empirically and dangerously, by ordinary

people. The use of LSD, for example, which until 1966 is legally produced and distributed in the United States, spreads like wildfire, especially among artists and intellectuals. In 1967, when its consumption is prohibited in the United States and, subsequently, in most countries in the world, a frenetic search begins for legal alternatives that have analogous capabilities. Homemade experimentation becomes more and more popular; in practice, any kind of substance, especially of a vegetable nature, is dried and smoked. The chronicles of the time are incredibly amusing. There are analytical reports and oodles of detailed descriptions of how to treat green and red peppers (acerbic, ripe, or rotten), chili peppers, eggplant, sage, oregano, potatoes, and any other vegetable or mushroom that you can think of that can be mixed with tobacco or smoked in pure form. In the midst of this frenzy, to the tune of an abstruse composition by the Scottish singer-songwriter Donovan, is where our story of banana powder begins.

In 1966, Donovan—known, with a heavy dose of optimism, as the "British answer" to Bob Dylan—releases a single called "Mellow Yellow," which quickly becomes a planetary hit, soaring to the top of the charts all over the world, including *Billboard*, where it reaches number two. It seems that it is precisely the total incomprehensibility of the words to the song, with their infinite possibilities of interpretation, which is the basis of the legend of the hallucinogenic banana. You can read anything into "Mellow Yellow," starting with the title. What in the world does mellow yellow mean? Pulpy yellow? Pasty yellow? Velvety yellow?

And if someone out there figures that the song's lyrics might be of some help in deciphering the title, they're wrong. If the title is ambiguous, the lyrics are totally obscure. The only phrases with a logical meaning talk, maybe, about a girl named Saffron and someone who refers to the singer as Mellow Yellow, whatever that might mean.

From obscure phrase to obscure phrase, we get, finally, to the incriminating verse: "Electrical banana / Is gonna be a sudden craze / Electrical banana / Is bound to be the very next phase." Onto this apparently nonsensical proclamation a story is grafted, according to which, Donovan is suggesting, with the amazing foresight that a prediction of such dimensions requires, that the future belongs to psychedelic trips fueled by banana skins. Forget marijuana, the future is bananas. In reality, Donovan's lyrics are the classic gibberish on display in lots of songs from that era: a sequence of phrases that mean absolutely nothing, whose only merit is that they match the rhythm of the music and seem to conceal an impenetrable mystery. Which is one reason, though certainly not the main one, why the comparison with Dylan goes down the tubes.

Asking the author, obviously, is useless. When asked, in decades of interviews, what exactly he was trying to say with his enigmatic banana, the most intelligible answer was that it was really a reference to a yellow vibrator that he saw in an advertisement. In any event, it was on the basis of this sibylline lyric that, in 1967, a group of young people in Berkeley, California, came up with the magnificent invention that banana skins, appropriately treated, are a potent hallucinogen. The times are certainly ripe. Remember that some of the most important albums of rock music released between 1965 and 1967 were inspired by drugs, including, just to cite a few, *Bringing It All Back Home* or *Highway 61 Revisited* by Bob Dylan, and the Beatles' *Rubber Soul* and *Revolver*. In those years, when rock brought psychedelic substances into popular culture, the conditions for the banana to be transformed into a mind-altering substance seemed to be perfect.

So, on March 3, 1967, the *Berkeley Barb*, one of the most influential underground magazines of the era, published a very detailed article by Ed Denson on the hallucinogenic properties of banana skins and the possible ways of obtaining a pleasant trip by smoking the equivalent of four cigarettes, appropriately spiced with the addition of potent banana powder. Denson recounts how he had gotten high on a banana joint just a few days before in Vancouver, where he had been "initiated" into the use of this new psychedelic substance, with amazing results. In the same issue of the magazine, a hilarious letter to the editor written by an anonymous "cooperative member and careful shopper" recounts having seen "two suspiciously off-beat gentlemen lurking in the fresh produce section" who were then revealed to be members of the Berkeley police's "narco squad." "I would guess that they have been assigned to observe persons purchasing large quantities of bananas," writes the anonymous cooperative member, who goes on to explain that it is by now well known that bananas contain psychoactive properties. The letter closes with the prediction that the possession of large quantities of bananas will soon be considered a crime.

Reading that letter today, it seems impossible not to realize that it was a joke. At the time, however, nobody seemed to get it, and the repercussions went far beyond all expectations. On March 4, the day after publication of the article in the *Barb*, the news is picked up and highlighted by the *San Francisco Chronicle*. The street-corner advertising posters for the paper report the news in large type. The article cites what was reported in the *Barb*, gives the name of the formerly anonymous careful shopper, and asks for a comment from the Berkeley police department, which denies any involvement. The article concludes with the opinion of a well-informed hippie, according to whom the whole thing was whipped up by the United Fruit Company[12] to sell more bananas.[13] If it really had been a publicity stunt, the results would have been fantastic: The next day, in the whole Bay Area, you couldn't find a single banana for sale.

The *Chronicle* article has an enormous circulation and is picked up immediately by dozens of other underground newspapers, which, in those years, recounted events in the various hippie communities scattered around the United States. From fanzines to local community underground papers, from letters to the editor to mimeographed flyers distributed at all the main hippie events, the news reaches the entire nation. The editors of *The East Village Other* filch the story, claiming they were the discoverers of the psychoactive capabilities of the banana; the *Los Angeles Free Press* circulates the recipe for the preparation of perfect Mellow Yellow, and everyone from coast to coast celebrates the powers of the banana skin powder. There are even advertisements for pioneering companies able to provide "100% pure legal bananas."

On March 16, 1967, the *Columbia Daily Spectator* publishes an electrifying article signed by Christopher Hartzell under the eloquent headline LOW-COST HIGH, which after a magnificent opening—"The grass is not always greener on the other side of the law, and a true psychedelic experience may be no farther than the nearest fruit store. The product is not marijuana, hash, LSD . . . but rather bananas—yes, plain old everyday yellow United Fruit Company style bananas"—gives directions for how to do it. "[O]ne can get high in a number of ways with bananas. The easiest is to peel a ripe banana, scrape from the inside of the peel the white fiber, and dry in a slow oven of about 200 degrees or less for about twenty minutes . . . The dried pulp can be crumpled to resemble tobacco, rolled into a cigarette, or smoked in a pipe. Another method is to place a stick of gum inside a sliced banana, and allow it two weeks rest before chewing."

The consumption of bananas rockets upward. For several weeks, in cities like New York and San Francisco, there are no more bananas on the counters of fruit and vegetable markets. Rumors spread that agents from narcotic squads are paying special attention to purchasers of unusually large quantities of bananas. In New York, during a three-day Cosmic Love-in at Central Park, improvised stands sell low-cost bananas from various countries, with detailed explanations of their mind-altering characteristics and of how to get the most punch out of the different varieties. Chants are sung in honor of the banana and a new greeting is invented, addressed to the numerous members of the happy and aware banana-consumer club. Basically, it is the classic raised middle finger but with a new wrinkle. Instead of straight up, the finger is slightly curved to recall the shape of, that's right, a banana. I imagine this led to more than a few misunderstandings. And then, obviously, banana skin powder is smoked everywhere. For a brief but picturesque period, doing a joint of banana powder is much more trendy than old-fashioned marijuana. The spread of the yellow joint, legal and still cheap, appears to be unstoppable.

On May 26, 1967, the Food and Drug Administration issues a press release in which it affirms that a laboratory study of several recipes for dried banana skins showed "no detectable quantities of known hallucinogens." Meanwhile, the hunt is on for a hallucinogen that some party-hardy freak has dubbed "bananadine." Although the findings of the FDA do not show any known psychedelic substance in bananas, nor anything else that can be described as bananadine, there is no slowdown in the alternative consumption of bananas, to the point that even universities decide they needed to study the problem.

The authoritative journal *Economic Botany* published an article signed by a Dr. Krikorian, which reports the correct recipe used to obtain the elusive banana powder.[14] Here it is:

Procedure:

(a) Procure 15 lbs. of ripe yellow bananas.

(b) Carefully peel all 15 lbs. and discard (or eat if hungry) the fruit. Save the peels.

(c) With a sharp knife, scrape off the insides of the peels and save the scraped material.

(d) Put the scraped material in a large pot(s), and add water. Boil for 2–3 hours until it attains a solid paste consistency.

(e) Spread the paste on cookie sheets and dry in the oven for about 20 minutes until the paste becomes a fine, black powder.

Makes one full pound of bananadine powder.

Although Dr. Krikorian was not able to identify any hallucinogenic substance in bananas, his conclusions nevertheless advise prudence. In fact, "Although the effects of banana smoke have been shown to be more psychological than psychedelic, the fact still remains that banana skins, like all other vegetable tissue, contain countless unidentified substances." Gradually, without any fanfare, unlike when it first appeared, the news fades away and vanishes from the newspapers. Every once in a while, the local news pages carry a story about a group of kids stopped by the narcotics squad for possession of pipes or aluminum foil containing Mellow Yellow, but this happens less and less often. In the fall of 1967, the underground papers, now occupied with the growing protest against the American involvement in Vietnam, the publication of *Black Power*,[15] and the spread of rioting in the Black urban ghettos, regrettably slough off their spirit of playfulness, allowing the banana to return, after a brief binge of transgression, to its ineluctable destiny as the stuff of baby food and smoothies.

Some of the pieces of wood used to build the ladder that was instrumental in the kidnapping of the son of Charles Lindberg came from a forest in South Carolina.

Seven

PLANTING LAW
AND ORDER

For years, in my youth, I was scared to death of all university assemblies or any other form of important academic meeting where I was required to participate. I'm not sure how much you know about how these conclaves work, but that doesn't matter. All that counts is to know that when a certain number of academics assemble, a ritual begins which follows precise rules that must be adhered to. The moment that afflicted me most was the conversation with my academic colleagues following the main speaker. In this case, too, the rules are clear and compulsory. First off, you introduce yourself, being very careful to display ample signs of respect upon hearing the name of your interlocutor. Then you shift to your comment on the presentation you have just listened to, mixing words of appreciation with a few, well chosen, polite criticisms of the speaker. It all has to be orchestrated in a way to make it understood that, although what you had listened to was an excellent presentation, both you and your interlocutor would have been able to do better. And then, right after the conclusion of this phase, there arrives the embarrassing moment I mentioned earlier.

In general, it works like this. Your interlocutor suddenly interrupts the continuous flow of words with which he has inundated you for the last few minutes and starts looking you intensely in the eye, as though he were trying to remember something that he knows perfectly well but that now, due to a lapse of memory, he can't seem to recall. After having thoroughly terrified you with ever more incomprehensible intense looks, when you have already started calculating the probability of your getting out of there alive, the fatal phrase is pronounced: "And you, Stefano, pardon me, can you remind me of just what it is that you do?" It is not at all easy to know how to articulate the question accompanied by the appropriate body language. You have to know how to enunciate it as though your colleague's occupation, universally known, escapes you in that moment, owing to a precocious attack of senile dementia.

It is exactly at this point that the ceremony shuts down. Indeed, if up to that precise point I had managed to pull it off fairly decently, the unveiling of my occupation, as incongruous as an onion in a cup of champagne, blocked the mechanism and jammed the works irreparably. "I study plants. I'm an agronomist," is a response not contemplated by the canon. Physician, physicist, chemist, jurist, architect, engineer are all respectable professionals, even philosophers, mathematicians, geographers, and geologists, though clearly eccentric, are recognized as members of academic disciplines. But an agronomist, come off it, what sort of profession is that supposed to be? The moment you say the word, you realize that you are placing yourself beyond decency.

At the start, your impeccable interlocutor takes it as a wisecrack. He can't believe that this distinguished gentleman, with whom he has been conversing amiably up to now, is an "agronomist." He doesn't even really know what exactly an agronomist is, only that it has something to do with the land. This bizarre word reminds him vaguely of great Russian literature. Wasn't it Tolstoy who mentioned, now and again, some agronomists in his novels? If he is a copious reader, he will pull out, with a forced smile, K., the protagonist of *The Castle* by Kafka. "Ah, interesting, so you do the same work that K. did?" You would like to tell him that K., in reality, is a land surveyor, and if we really want to get right down to it, in a certain sense, *The Castle* is constructed entirely on the misunderstanding of K.'s profession. But you never have enough time. Once they know you're an agronomist, they all run away, like frightened fawns.

I believe this is a common experience of anyone who works with plants. Those who study them share with the plants they study their complete insignificance. Research in the vegetable realm is considered to be something that belongs more to the world of farming than to the world of science. Plant researchers are eccentric gentlemen and ladies who don't have anything better to do than occupy themselves with this marginal green stuff, instead of devoting themselves to more serious and weighty research, as true scientists do. One consequence of this distorted view of reality is that in all the fields of knowledge, from cellular biology to anatomy, from ecology to the history of evolution, the discoveries attained thanks to plants always have an importance approximating to zero when compared to those attained in the animal world. This explains why—and I've written about this previously—many fundamental discoveries regarding, for example, cell biology, after being completely ignored in their experimentation in the vegetable realm, have been rewarded with Nobel Prizes for those who have slavishly replicated the experiments in some insignificant animal organism. It's as though something that is valid for the 0.3 percent of life represented by animals were of greater importance than what is valid for the 85 percent of life represented by plants. I'll never be able to understand it.

A prime example is the centuries-long marginality of plants as a source of legal evidence. How could it have been possible to ignore the omnipresence of plants? In reality, we have only botany to thank for the chance to solve some of the past's most sensational court cases. In fact, it was actually the resolution of the so-called crime of the century, namely the kidnapping of the firstborn child of the celebrated aviator Charles Lindbergh, that, for the first time in history, fundamental evidence of a botanical nature led to the identification of the culprit and was admitted into evidence during the trial. That is the story and it deserves to be told in its entirety.

In 1927, Charles Lindbergh, a twenty-five-year-old pilot, became immensely popular the world over thanks to his epic exploit of flying solo across the Atlantic from New York to Paris aboard the *Spirit of St. Louis*, the plane that he himself had helped design and build. In the eight years since 1919, when Raymond Orteig, a wealthy hotel owner from New York, had instituted a $25,000 prize, named after himself, for the first pilot to fly nonstop between New York and Paris, only a very few had accepted the challenge. And, for those few, the results had not been encouraging in the least. The main problem to be resolved was how to transport the fuel necessary to complete the crossing.

In 1926, the French ace René Fonck, behind the controls of a plane hopelessly overloaded with fuel, crashed on takeoff, losing two crew members in the accident. In 1927, the year of Lindbergh's crossing, three different American crews and a French one had already failed, paying a heavy tribute of deaths, before Lindbergh, an unknown postal service pilot with very few resources at his disposal, succeeded in the enterprise.

Lindbergh bets everything on the lightness of his plane. First, the *Spirit of St. Louis* is a single-engine, single-wing design, unlike the double-wing bi- or tri-motor models which, up to then, had attempted the flight. Moreover, in order to reduce the plane's weight further, he decides to make the crossing by himself, with no crew, and to eliminate from the plane any object or instrument (radio included) that he judges to be not strictly necessary. Finally, to resolve the fuel problem, he redesigns the nose of the airplane so it can hold an enlarged fuel tank. This modification makes any forward vision impossible. But Lindbergh decides that it's enough to be able to rely on the little side windows and on a sort of rudimentary mirrored periscope that allows him to see ahead.

In these conditions, at the helm of an airplane transformed into an enormous flying fuel tank and without any forward vision, Lindbergh takes off at 7:52 on the morning of May 20, 1927, from Roosevelt Field, near New York. His legendary flight of thirty-three hours and thirty-two minutes will bring him to land at Le Bourget, on the outskirts of Paris. From one day to the next, Lindbergh becomes one of the most famous people on earth. In the United States, he is appointed colonel in the air corps reserve and awarded the Distinguished Flying Cross. The French government awards him the Legion of Honor and *Time* names him Man of the Year.

Eight years after this successful enterprise, on the evening of March 1, 1932, when his firstborn child, Charles Augustus Lindbergh Jr., just twenty months old, is kidnapped from the family home in New Jersey, Lindbergh is still one of America's idols, and the news creates a huge turmoil. The circumstances of the kidnapping are described in every newspaper. After dinner, the governess, Betty Gow, puts the baby to bed in his crib in a room adjacent to the library, where his father is located at the time. Around nine thirty, Lindbergh hears a noise in the house but he pays no attention, assuming it was caused by some object in the kitchen falling on the floor. At ten o'clock, the governess looks into the baby's room to check on how he is sleeping and finds the crib empty, the window wide open. Once it is ascertained that the baby is not with his mother, it is immediately made clear that the family is facing a kidnapping, when they find on the windowsill in the baby's room an envelope with a ransom request. Lindbergh calls the police and, in the meantime, he searches the yard, finding some tire prints and a ladder with wooden rungs hidden in the bushes. In the months to follow, this very same ladder will turn out to be one of the principal pieces of evidence for the prosecution. But let's take it one step at a time.

The investigation is conducted by the superintendent of the New Jersey State Police, Herbert Norman Schwarzkopf Sr., whose son, "Stormin' Norman," will lead the multinational allied forces deployed in the First Gulf War sixty years later. For months the investigation yields no results, notwithstanding payment of the ransom request. On May 12, 1932, a truck driver, having stopped to urinate in the woods along a country road a few miles from the Lindbergh estate, finds the child's body.

The following is a summary, quite condensed, of the events that, two years later, led to the identification of the culprit and to the hearing in which, for the first time, botanical evidence was admitted in a criminal trial. But let's not get ahead of the events and, instead, follow the operations of the police.

The turning point in the investigation comes when the police discover some information from the banknotes used to pay the ransom, whose serial numbers had been carefully transcribed.[1] Having noted that most of the incriminating bills had been spent along the route of the IRT Lexington Avenue subway line in Manhattan, the investigators concentrated their efforts on checking with the banks and the main commercial establishments within a few hundred yards of the subway. An operation that took a long time but that, thanks to a fairly large dose of luck, led to the hoped-for results. One of the ten-dollar bills, identified by the Corn Exchange Bank, carried a penciled notation on the edge of the bill of a license plate number: 4U-13-41-NY. The detectives managed to work their way back to the owner of a gas station, who had deposited the bill in his bank account. He recounted that he had noted the license plate number of the vehicle (a blue Dodge), whose driver had asked him for a full tank of gas, because he suspected that the bill might be counterfeit. The owner of the car turned out to be Bruno Richard Hauptmann, a German immigrant who worked as a carpenter, with a criminal record back in his homeland. His address was 1279 East 222nd Street in the Bronx. In his garage, agents found $14,000, which was part of the ransom.

Although this was serious evidence of guilt, the presence of the money, by itself, would not have been enough to incriminate Hauptmann. The money could have been placed in his garage by someone else or it could have been handed over to him by the real kidnapper. In short, this was important evidence, but not the so-called smoking gun that any jury would recognize as undeniable proof of guilt. The search had to continue for something that unequivocally tied the suspect to the house in New Jersey, where the Lindbergh baby had been kidnapped. Convinced by now that they were on the right road, the investigators carried out a meticulous search of Hauptmann's home. They found more evidence: a notebook with sketches of a ladder very similar to the one found in the Lindberghs' yard right after the kidnapping and, most important, a piece of lumber from which there appeared to have been sawed a piece of wood similar to the one used to make a side rail of the ladder used in the kidnapping.

Similar, however, does not mean identical. For the evidence to be considered valid, every doubt that the side rail came from the same piece of wood found in Hauptmann's home had to be eliminated. This is where our hero comes in: Dr. Arthur Koehler, a talented expert in anatomy and the identification of wood at the Forest Products Laboratory of the Forest Service in Madison, Wisconsin. It is solely thanks to his work that it was possible to solve one of the most famous cases in American legal history. Right from the start of the investigation, charged with examining the infamous ladder left by the kidnapper in the Lindberghs' backyard, Koehler realized that a detailed examination of the ladder could lead to precious information that would help identify the kidnapper.[2] At the same time, it was exceedingly clear to him that the validity of the eventual evidence found following his study of the wood would never be admitted at trial, unless it was glaringly certain. Never before, in fact, had evidence of a botanical nature been admitted in a court of law. For Koehler to have some chance of being heard, he had to find evidence in the wood of the ladder that left no room for doubt . . . as well as, naturally, an intelligent trial judge.

Now, to follow the rest of the story, it is necessary to know that wood was one of the first things examined by Antoni van

Leeuwenhoek, the pioneer of microscopy, at the beginning of the seventeenth century. Since then, observation of the anatomy of tree trunks has fascinated anyone who has a passion for the work. What do these observations tell us? That a trunk is essentially a hydraulic structure. An uninterrupted series of hollow dead cells arrayed to form long conduits that transport the water, and the solutes dissolved in it, from the root to the leaves. Along with the parenchyma, in which they are enclosed, these structures form the xylem (from the Greek *xylon*, wood). On the contrary, the living cells of the phloem (from the Greek *folios*, bark) transport the sugars produced in the leaves from the crown to the rest of the tree, thanks to photosynthesis. Each species tends to distribute the various types of cells that form the xylem in distinctive arrangements. So every good botanical anatomist is able to identify the species that produced a certain type of wood.

This is exactly what Koehler set out to do. He took the ladder to his laboratory and began to examine it in detail. In no time, he understood a fundamental fact: that the ladder was homemade, that is, it was not a serially manufactured object, which could be found in thousands of identical items, but a unique object, which one could expect would reveal important information about the person who had made it. In effect, the poor workmanship and imprecision with which the ladder had been constructed indicated that its maker, albeit possessed of some rudimentary know-how, was surely not a capable carpenter. Moreover, the rungs, even though they were made with a very soft wood such as that of the *Pinus ponderosa*, looked brand-new, without any signs of wear, demonstrating that the ladder had never previously been used and was, therefore, built for that specific purpose.

The wood used to make the ladder belonged to four different species. Three of the six rails (it was a telescopic ladder divided into three distinct sections) were made from *Pinus echinata* or related species, while the other three were from *Pseudotsuga menziesii*, as was one of the eleven rungs. The other ten were made from *Pinus ponderosa*. The hinges used to hold together the three sections were made from birch (probably *Betula alba*). Finally, the famous rail 16, crucial for the definitive conviction of Hauptmann, presented some peculiarities that did not escape Koehler. First, four nail holes indicated a previous use of the wood, since they had no function in the construction of the ladder. Then, the state of conservation of the wood, which, still luminous and with no signs of rust around the nail holes, indicated it had been conserved in a closed space, without any exposure to the elements.

Based on his observations and considering the poor quality of the lumber, Koehler suggested that the piece of wood used for side rail 16 came from a barn, a garage, or an attic. In addition to the fundamental deductions regarding side rail 16, which, as we shall see, will nail Hauptmann as the culprit, Koehler demonstrated his extraordinary expertise in many other phases of the investigation. For example, by succeeding in identifying, from the marks left by the plane on some parts of the ladder, the sawmill in South Carolina where the lumber for the side rail was cut. Then, by discovering the retail store of the National Lumber and Millwork Company in the Bronx, just ten blocks from Hauptmann's home, where the lumber for the side rail was sold.

But all of this happened long before the police arrested the suspect and found that one of the boards in the floor of his attic had been partially cut and removed. Once they'd found the board, as foreseen by Koehler, the smoking gun was within reach. If Koehler were able to demonstrate, beyond any doubt, that side rail 16 of the ladder used to kidnap Charles Lindbergh Jr. came from the cut board in Hauptmann's attic, no jury in the world would have any more doubts. That's exactly what happened. By studying the pattern of the annual growth rings in the attic floorboard and in side rail 16, Koehler ascertained that the rail came from precisely that board in Hauptmann's attic floor. Not only that. To dispel any doubt whatsoever, Koehler demonstrated that the concentric rings of the wood, just like fingerprints, have unique markings, such that no two trunks could ever exhibit identical patterns.

In January 1935, in a climate of public interest and involvement without precedent in American judicial history, Koehler was called to testify on the fifth day of the trial and, in a much more detailed way, as a key witness right before the final arguments. At that point, the fundamental importance of Koehler's testimony was clear to everybody, especially to the attorney for the defense, Edward J. Reilly, who tried in every possible way to discredit the testimony, appealing to the "fact" that the study of wood "*was not a science.*" In fairness, it must be recognized that, at the time, the use of expert witnesses in criminal trials was, in general, rare and limited. Nevertheless, despite this unfavorable custom, experts in disciplines different from botany were sometimes called.

It's interesting to recall the exact and derisive words with which Reilly addressed the court:

> *There does not exist among men an animal known as an expert on wood; which is not a science recognized by the courts; which has nothing to do with handwriting experts, finger print experts, or ballistics experts. Those are some of the sciences recognized by the courts. This witness can probably testify as an expert carpenter or something like that, but when he tries to qualify himself as an expert or express opinions as an expert on wood, that is something rather different ... he is simply a man who has a lot of experience examining trees, who knows about tree bark and a few other things of that kind. He can come into court and tell us what he did and what he saw, but when it comes to expressing his opinion as an expert or a scientist, that is very different. Let's say that the opinion of the jurors is just as good as his and they are every bit as qualified as he is to judge.[3]*

Fortunately, the judge did not see it that way, and by pronouncing the phrase, "I say this testimony can be qualified as expert,"[4] he signaled, for all practical purposes, Bruno Hauptmann's sentence to death in the electric chair and the birth of forensic botany.

Despite these illustrious origins, and even though almost ninety years have gone by since the Hauptmann trial, botany, not only in the forensic sphere, continues to be the poor relation among the scientific disciplines. Inexplicably. Indeed, vegetable remains of the most various kinds are present everywhere. It is not possible to be devoid of vegetable matter. In other words, it is unimaginable that people, who represent a negligible fraction of the planet's biomass, do not retain a trace of that decisive 85 percent of vegetable biomass in which we are all immersed. Vegetable remains are literally all over the place and offer multiple potential sources of evidence, both at the macroscopic level (wood, also in burned form, leaves, fruit, twigs, flowers, roots, etc.) and at the microscopic level (hairs or trichomes, algae, spores, pollen, etc.). The morphological diversity of every possible vegetable specimen, the association and comparison of different species, and their relative quantities allow us to gather fundamental information for understanding, for example, the season or the geographical position in which a crime was committed, independently of the fact that a body has been moved or buried. Vegetable matter found on the body or clothing of a suspect can tell us about his or her presence or not at the scene of the crime. Pollen and spores, in particular, being highly diffuse in the air, not only attach themselves in significant quantities to our clothing but are also inhaled and can be identified in our airways.

Palynology (an obscure branch of botany that studies the pollen of other microscopic biological elements such as moss spores, lycopods and ferns, spores and fungal parts, both contemporary and fossil) is one of the most prolific scientific disciplines for significant information in the solution of judicial cases. The most famous, and also the first, case solved thanks to palynological analysis dates back to 1959 in Austria, and involves the sudden disappearance of a man who was presumed murdered despite his body never having been found. The investigators had a suspect but very little evidence, including a pair of boots splattered with mud. A local palynologist analyzed them and found something decidedly uncommon: the fossilized pollen of a walnut tree from twenty million years ago. That kind of tree had not grown in Austria for millions of years, but its fossilized pollen could still be found in a small region along the Danube. Armed with these findings, the investigators succeeded in making the suspect confess the location of the corpse.[5]

Palyonology and the study of pollen have made it possible to solve war crimes in East Germany,[6] and to figure out the last movements of Ötzi, the Ice Man of Similaun Glacier, and the time of the year in which he died. Indeed, the growth of roots over buried bodies and objects can give us, through the study of their growth rings, important information on the date of burial.[7]

Another source of fundamental evidence and data is, obviously, the molecular analyses of DNA. This was true for the case of a girl whose murderer deposited her body in the middle of the Arizona desert. It was May 2, 1992, and thanks to the work of Tim Helentjaris of the University of Arizona, the investigators were able to ascertain that two pods found in the van of one of the suspects came from a tree of the species *Parkinsonia aculeate*, present at the scene of the crime. This was the first time in history that a court of law admitted evidence coming from vegetable DNA.

Despite all of this, if you look at the scientific disciplines represented in forensics laboratories, it would seem that even today plant experts are not contemplated. The personnel of the Scientific Investigation Division of the Carabinieri in Italy, for example, includes specialists in molecular biology (mainly for the analysis of DNA and organic traces); chemistry (nonbiological traces such as fibers, paint chips, and unidentified liquids and chemical substances); ballistics (everything that concerns weapons, firearms, knives, swords, bayonets, etc.); dactyloscopy (fingerprint analysis); phonics and graphics (voice and graphological comparisons, anthropometric exams); psychiatry and psychology (elaboration of psychological profiles in the most heinous crimes without an apparent motive); and sociology, criminology, statistics, and information technology (studies and research on persecutory acts—stalking—and manifestations of violence and abuse of vulnerable victims). But there is no mention of botanists or plant experts at any level.

Analysis of the professional skills required by the other principal scientific investigation departments around the world shows pretty much the same situation. A few forensic laboratories have entomologists on staff, but never a botanist. In the most recent available version of its *Handbook of Forensic Services*, the FBI mentions wood fibers but makes no reference to any other type of vegetable matter samples that could be taken from the crime scene and analyzed in the laboratory.[8] The entire handbook has no mention, for example, of the word "pollen." No one should be surprised by this. In a 1990 survey of the thirty largest forensic laboratories in the United States, only two were aware that pollen could be used as a forensic tool.[9] The objection raised by Hauptmann's defense attorney that botany "is not a science" is still alive and well in the common wisdom.

Plants were the first living beings to travel in outer space and they will always be with us when we explore the solar system.

Eight

PLANTING THE MOON

A t a time when plants had already established their place in my life, I was distractedly leafing through an American gardening magazine, when I was suddenly blown away by an advertisement trumpeting the chance to have in your very own garden the same trees that important historical figures had loved, for whatever reason. The ad that had so irresistibly attracted my attention showed a gentleman dressed in eighteenth-century garb, complete with a walking stick, contently at ease under the shade of an enormous tree. The slogan accompanying the image was "Walk with Washington under his favorite tree." For sixty dollars, plus shipping, you could receive from the United States a small tree born by way of vegetative propagation (a clone) from a tree related to key moments or figures in the history of America.

Coming to know of this opportunity filled me with infantile joy. Practically speaking, it was now possible to take delivery of the red maple that had inspired Henry David Thoreau, or Mark Twain's oak, or even the sycamore trees planted by George Washington at Mount Vernon in 1795, and so on. And we were talking about exactly *the very same tree*. Sure, not in the dimensions and in the place where they had performed their historical function, but nonetheless the same trees. Transferring this opportunity into the animal kingdom, it's as though we could have Bucephalus, the horse of Alexander the Great, with us again, or Marengo, the horse that carried Napoleon to Marengo, Jena, Wagram, and Waterloo, or the dog Laika, the first living being to travel in space.

The offer was undeniably charming, and for a long time the list of the various available trees was a topic of discussion with my collaborators in Florence. Should we order Washington's sycamore or Thoreau's maple? Each of us had favorites. And why not create a buyers club and multiply the trees on our own account so that we could all have one of each? A little like when we were kids and pooled our records. Each of us was the owner of the record they had chosen and paid for, the others were free to copy it onto a cassette. Except that, in the case of the trees, the system would have worked even better. Unlike what happened with the cassette copies, the cloned trees would have been identical to the original.

As with the records, however, the discussions were endless and, in the end, demanded judgments on the merits and painful choices, and for what seemed like forever, we talked about nothing else in our laboratory. Then, just when everything appeared to have been decided and the final laboriously generated list was ready, there appeared on the horizon the real mortal enemy: bureaucracy. It became immediately clear to us that, not being a well-known plant nursery, importing plants to Italy from the United States was something beyond our reach. Certainly well beyond mine. Just the sight of the list of things to do and certifications to produce had the usual paralyzing effect on me that, by now, bureaucracy succeeds in inducing for procedures much less onerous. I am convinced that this is the ultimate objective of bureaucracy: to block any and all requests. Well, in my case, the objective has been achieved. The bureaucracy has won: I will never make another request. Every now and again, I get the idea that there is a secret office of Security Services for the Defense of Bureaucracy, which fights behind the scenes to attain this objective. I imagine an office, whose staff, every time a citizen renounces forever the right to request something from public institutions, drinks a toast to the mission accomplished and adds the citizen's name to the golden list of the Silent: the list of perfect citizens.

As often happens, however—and this is something bureaucracy has not managed to block—knowledge still generates possibilities, and in this case, too, the unforeseeable happened. My learning of the catalogue of historic trees enkindled my passion for famous trees and, among them, the little-known but misunderstood moon trees. It was an outright miracle of which I am about to tell you the story.

On February 5, 1971, the Apollo 14 mission lifted off from Cape Kennedy (now again named Cape Canaveral), destined for the moon. The crew was composed of Alan Shepard, the commander; Stuart Roosa, the pilot of the command module; and Edgar Mitchell, the pilot of the lunar module. This was the third human lunar landing mission and, for the occasion, Alan Shepard, a pioneer of space exploration, was asked to take command of the mission and be the fifth man (of the twelve who have ever done it) to set foot on the lunar surface. Actually, Shepard had been the second human and the first American to go into space. That happened on May 5, 1961, less than a month after the legendary voyage of Yuri Gagarin on April 12, 1961, demonstrating to the Soviet Union that the technological gap between the two superpowers in the field of space exploration was not insurmountable.

The popular image of Shepard is of the classic American hero: aviator, naval officer, astronaut, able to go back into space decades after his first mission and to take with him to the moon two golf balls with which, using a six iron, he hit a couple of shots at reduced gravity that, as he described them, flew for "miles and miles." In short, Shepard remains, still today, the only moon golfer, but he is no longer the player to hit the longest "golf shot." In the rankings of this very special competition, he has been surpassed by the cosmonaut Mikhail Tjurin, who in 2006, hit a golf ball off the International Space Station, sending it in the direction of the earth's atmosphere.

In any event, we are not talking about golf here and the hero of our story is not Shepard. The hero is another astronaut of the crew on the same mission: Stuart Roosa, the commander of the lunar module during the Apollo 14 mission. Roosa had a past every bit as adventurous as Shepard's, but in addition—and the reason why he is by far my favorite astronaut—he had a passion for trees and forests. Before joining the space program, in fact, through the first half of the 1950s, Roosa had combined his two main interests—adventure and flight—in one of the least known and most dangerous jobs that you can imagine: smoke jumping. What is it? Simply put, it is parachuting into forests situated in impervious territories where a fire has started in order to extinguish the fires before they get out of control. The parachuters dig trenches and execute all the maneuvers that can slow down the spread of the flames, while waiting for the arrival of other firefighting equipment and personnel. From Roosa's accounts, it appears that his love for trees came directly from his close encounters with their crowns, in those countless times when his parachute got entangled in their branches, leaving him suspended way above the ground. While swinging back and forth among the foliage and trying to free himself without breaking his neck, he had learned to appreciate these enormous, marvelous beings. So when, just a few days before liftoff, Ed Cliff, the chief of the Forest Service, called Roosa to ask if he would be willing to carry on board of the Apollo 14 spacecraft a metal container filled with five hundred seeds, he said yes, without any hesitation.

The seeds in question belonged to numerous species commonly found in the United States, such as liquidambar, sequoias, Douglas firs (*Pseudotsuga menziesii*), sycamore, pine, etc. The choice fell to these species both because they were very well known and because they would be able to thrive in most regions of the United States. The idea behind the request made of Roosa was to be able to follow the growth of the seeds that had been in outer space, comparing it to the growth of seeds produced by the same parents but which had never left the earth. Thus, for every seed sent into space, some of its siblings were set aside. In other words, the idea was to have trees growing near each other, some of which had been to space and some of which had not, so as to be able to assess the eventual differences. None of this came about. In fact, when Apollo 14 landed in the southern Pacific on February 9, 1971, the seeds risked being destroyed. During the decontamination process, the container, disinfected and depressurized, burst open, scattering the seeds all around and vastly reducing the chances they would ever germinate.

The first attempts by the NASA staff to get those seeds to germinate came to nothing, but more for incompetence than for anything else. Entrusted to more skilled and patient hands, almost all the seeds germinated, giving life to several hundred small trees that had lived through the adventure of being to the moon. In 1976, many of the trees born from those special seeds were planted around the United States, to celebrate the bicentennial of the Declaration of Independence. The idea was that trees that had been to the moon were in harmony with the spirit of the Founding Fathers. A pine tree was planted in the garden of the White House in the presence of President Gerald Ford, who described these trees as "the living symbols of our spectacular human and scientific achievements." A sycamore was planted in Washington Square in Philadelphia, another near the Kennedy Space Center in Cape Canaveral, a sequoia in Berkeley, a Douglas fir in front of the smoke-jumpers base in Oregon, and so on. Many of the moon trees were planted in front of schools, universities, courthouses, and public offices all over the United States. Some of them went abroad, to Brazil, Switzerland, and Italy. One of the trees was presented to Emperor Hirohito of Japan. Then, just as quickly as it arose, the great interest in these trees disappeared and they were completely forgotten.

And they would have remained forgotten if it were not for the NASA archivist Dave Williams, who in 1996 received a strange e-mail from Mrs. Joan Goble, a teacher in Cannelton, a minuscule town in Indiana, asking for information about a tree growing near the Camp Koch Girl Scout Camp in Cannelton. Mrs. Goble thought NASA might know something about the tree because there was a plaque next to it, classifying it as a moon tree. Nobody in town seemed to remember anything special about that particular tree and the teacher, believing it unlikely that the tree had come from the moon, a place notoriously empty of life, wanted to ask NASA to shed some light on the mystery.

The story of this moon tree was totally new to Williams. Although his job title was archivist and he specialized in the Apollo missions, he had never heard anything about them. And

STEFANO MANCUSO

like him, practically everyone else in the NASA archives had never heard anything about them. It all could have ended there, with that bizarre e-mail, if it were not for Williams's stubbornness. As a good archivist, he did not let himself be discouraged by the initial difficulties but decided to look for information both from the few people still in service at NASA from the 1970s and from the Forest Service.

That's how the story of the moon trees came back to light. That there was nothing about them in the NASA archives was a consequence of the fact that the seed container was not part of an official experiment of the space agency but rather an individual initiative of a single astronaut. In fact, Roosa, instead of carrying in his personal baggage dozens of knickknacks to sell afterward at high prices to collectors of objects connected to space flights, had meritoriously decided to invest a large part of his weight allowance in the transport of a can filled with seeds. Williams succeeded in unveiling the secret. Now he knew enough to respond exhaustively to Mrs. Goble's e-mail, but not enough to satisfy his own curiosity about the whole story. How many of these trees were there? Where had they been planted? And, above all, how many of them were still living? And in what conditions?

Williams just could not let it end there. He had the impression that by forgetting those trees, NASA was committing an injustice toward living organisms that, no matter how you look at it, had been aboard Apollo 14 and had orbited the moon thirty-four times, only to be forgotten. He did not give up. By researching the newspapers of the 1970s, he succeed in locating a good number of

194

the trees, including the one planted in the White House garden by President Ford, and a sycamore planted right outside of his own office at the Goddard Space Flight Center in Maryland, whose status as a moon tree no one remembered anymore. Plus, he created a Web page, where he asked anyone with information about the moon trees to register them. Following this initiative, the press once again, for a brief period, became interested in the story and rekindled public interest in the trees. In the end, Williams was able to put together a list of seventy or so of these space-flight veterans, whose main characteristics (species, date of planting, location, and conditions) are still available to everyone on the NASA website, as is only right for these fine spaceflight companions.

NOTES

PROLOGUE

1. Edward Elgar, *Letters of a Lifetime* (Oxford: Oxford University Press, 1991).

1 PLANTING LIBERTY

1. This was a pun on the name of the Count of Bute, whose English pronunciation is the same as "boot," while the green of the soil was a play on Lord Grenville.

2. In scientific representations, graphs are formed by a set of elements called nodes (vertices) or peaks connected by lines called arcs or sides or edges. In the case of the fraternity trees, the image is a root network in which the nodes represent localities where liberty trees were planted and the arcs are the roots that connect those localities.

2 PLANTING CITIES

1. Arnold J. Toynbee, *Cities on the Move* (Oxford: Oxford University Press, 1970).

2. United Nations, *World Urbanization Prospects: The 2018 Revision* (New York: United Nations, 2019).

3. This is according to Columbia University Socioeconomic Data and Applications Center Gridded Population of the World and the Global Rural-Urban Mapping Project (GRUMP).

4. One orbit around the earth at an altitude between the atmosphere and the Van Allen belt, or between 100 and 1,250 miles.

5. Marcel Cardillo et al., "The Predictability of Extinction: Biological and External Correlates of Decline in Mammals," *Proceedings of the Royal Society B* 275 (2008): 1441–448.

6. Martin Warren et al., "Rapid Responses of British Butterflies to Opposing Forces of Climate and Habitat Change," *Nature* 414, no. 6859 (2001): 65–69.

7. Colin R. Townsend, Michael Begon, and John L. Harper, *Essentials of Ecology* (Oxford: Blackwell, 2nd edition, 2003).

8. Joseph Grinnell, "The Niche-Relationships of the California Thrasher," *The Auk* 34, no. 4 (1917): 427–33.

9. Edward Glaeser, "Cities, Productivity, and Quality of Life," *Science* 333 (2011): 592–94.

10. David E. Bloom, David Canning, and Günther Fink, "Urbanization and the Wealth of Nations," *Science* 319 (2008): 772–75.

11. Edward Glaeser, *The Triumph of the City: How Our Greatest Invention Makes Us Richer, Smarter, Greener, Healthier, and Happier* (New York: Penguin Books, 2012).

12. Mathis Wackernagel and William E. Rees, *Our Ecological Footprint: Reducing Human Impact on the Earth*, illustrated by Phil Testemale (Gabriola Island, Canada: New Society Publishers, 1996).

13. Alan Calcott and Jamie Bull, "Ecological Footprint of British City Residents," WWF report, available at assets.wwf.org.uk/download/city__footprint2.pdf, October 2007.

14. Hannah Ritchie and Max Roser, *Land Use*, published online at Our-WorldInData.org, available at https://ourworldindata.org/land-use, September 2019.

15. Erle C. Ellis et al., "Anthropogenic Transformation of the Biomes, 1700 to 2000," *Global Ecology and Biogeography* 19, no. 5 (2010): 589–606.

16. *The State of the World's Forests 2012* (Rome: Food and Agriculture Organization of the United Nations, 2012).

17. Joseph Poore and Thomas Nemecek, "Reducing Food's Environmental Impacts Through Producers and Consumers," *Science* 360, no. 6392 (2018): 987–92.

18. Robert A. Beauregard, "Planning with Things," 24th Annual AESOP Conference (Helsinki, July 7–10, 2010).

19. In this regard, it seems important to recall that the near totality of human innovations have been conceived and developed in urban environments.

20. Patrick Geddes, *Chapters in Modern Botany* (London: John Murray, 1911).

21. Patrick Geddes, "On the Coalescence of Amoeboid Cells into Plasmodia, and on the So-Called Coagulation of Invertebrate Fluids," *Proceedings of the Royal Society of London* 30 (1879–1880): 252–55.

22. Patrick Geddes, *Cities in Evolution: An Introduction to the Town Planning Movement and the Study of Civics* (London: Williams & Norgate, 1915).

23. Tom Matthews, Robert Wilby, and Conor Murphy, "Communicating the Deadly Consequences of Global Warming for Human Heat Stress," *PNAS* 114, no. 15 (2017): 3861–866.

24. Patrick E. Phelan et al., "Urban Heat Island: Mechanisms, Implications, and Possible Remedies," *Annual Review of Environment and Resources* 40, no. 1 (2015): 285–307.

25. Timothy R. Oke, "The Energetic Basis of the Urban Heat Island," *Quarterly Journal of the Royal Meteorological Society* 108, no. 455 (1982): 1–24.

26. Lisa Zaval and James F. M. Cornwell, "Effective Education and Communication Strategies to Promote Environmental Engagement," *European Journal of Education* 52 (2017): 477–86.

27. Jean-François Bastin et al., "Correction: Understanding Climate Change from a Global Analysis of City Analogues," *PLOS One* 14, no. 10 (2019): e0224120.

28. Jean-François Bastin et al., "The Global Tree Restoration Potential," *Science* 365 (2019): 76–79.

3 PLANTING THE UNDERGROUND

1. Martin Bader and Sebastian Leuzinger, "Hydraulic Coupling of a Leafless Kauri Tree Remnant to Conspecific Hosts," *Science* 1–6 (2019), available at https://doi.org/10.1016/j.isci. 2019.05.009.

2. René Joachim Henri Dutrochet, "Observations sur la longue persistance de la vie et de l'accroissement dans la racine et dans la souche du *Pinus picea L.*, après qu'il a été abattu," *Annales des Sciences Naturelles* 29 (1833): 300–3.

3. Barry Graham and Frederick Herbert Bormann, "Natural Root Grafts," *The Botanical Review* 32 (1966): 255–92.

4. I have written about this in my book *The Revolutionary Genius of Plants* (New York: Atria Books, 2018).

5. The vigor and proximity of a tree can be measured by one of its neighbors through the shade it produces and the volatile organic compounds they emit. See, for example, Ragan M. Callaway, "The Detection of Neighbors by Plants," *Trends in Ecology & Evolution* 17 (2002): 104–5, and Wouter Kegge and Ronald Pierik, "Biogenic Volatile Organic Compounds and Plant Competition," *Trends in Plant Science* 15 (2010): 126–32.

6. Jon E. Keeley, "Population Variation in Root Grafting and a Hypothesis," *Oikos* 52 (1988): 364–66.

7. Khadga Basnet et al., "Ecological Consequences of Root Grafting in Tabonuco (*Dacryodes excelsa*) Trees in the Loquillo Experimental Forest, Puerto Rico," *Biotropica* 25 (1993): 28–35.

4   PLANTING MUSIC

1.   Gherardo Chirici et al., "Assessing Windthrow Damage Using Single-Date, Post-Event Airborne Scanning Data," *Forestry* 91 (2018): 27–37.

2.   Barru Gardiner et al., "Destructive Storms in European Forests: Past and Forthcoming Impacts," final report to European Commission Directorate-General for Environment, 2010, available at https://ec.europa.eu/environment/forests/pdf/STORMS%Final-Report.pdf.

3.   The name of the genus *Picea* derives from the Latin *pix picis*, "pitch," referring to the abundant production of resin by many species of this genus.

4.   Lloyd Burckle and Henri D. Grissino-Mayer, "Stradivari Violins, Tree Rings, and the Maunder Minimum: A Hypothesis," *Dendrochronologia* 21 (2003): 41–45.

5.   Peter Ratcliff, "Violin Detective," *Nature* 513 (2014): 486.

5   PLANTING TIME

1.   Leonardo da Vinci, *Code Vatican Urbinate 1270* (Rome: Pontifical Lateran University Library).

2.   Mars is said to be in opposition to the sun when Mars and the sun are aligned with the earth but on opposite sides of it. In this configuration, Mars is, along with Venus, the brightest celestial body and is in the best conditions for being observed from earth.

3.   Stephen E. Nash, *Time, Trees, and Prehistory: Tree-Ring Dating and the Development of North American Archaeology 1914–1950* (Salt Lake City: University of Utah Press, 1999).

4.   Harold C. Fritts, *Tree Rings and Climate* (Caldwell, NJ: The Blackburn Press, 1976).

5.   Andrew Ellicott Douglass, "The Secret of the Southwest Solved by Talkative Tree Rings," *National Geographic Magazine* 56, no. 6 (1929): 736–70.

6.   Andrew Ellicott Douglass, *Climatic Cycles and Tree-Growth*, vol. III (Washington, DC: Carnegie Institution, 1936), 171 et seq.

7.   Donald J. McGraw, *Edmund Schulman and the "Living Ruins": Bristlecone Pines, Tree Rings, and Radiocarbon Dating* (Bishop, CA: Community Printing and Publishing, 2007).

8.   Edmund Schuman, "Longevity Under Adversity in Conifers," *Science* 119, no. 3091 (1954): 396–99.

9.   Edmund Schuman, "Bristlecone Pines, Oldest Known Living Thing," *National Geographic* 113 (1958): 354–72.

10. E. C. Anderson et al., "Radiocarbon from Cosmic Radiation," *Science* 105 (1947): 576.

11. J. R. Arnold and Willard Frank Libby, "Age Determinations by Radiocarbon Content: Checks with Samples of Known Age," *Science* 110, no. 2869 (1949): 678–80.

12. Willard Frank Libby, *Radiocarbon Dating* (Chicago: University of Chicago Press, 1952).

13. Available at https://www.nobelprize.org/prizes/chemistry/1960/ceremony-speech/.

14. Collin Renfrew, *Before Civilization: The Radiocarbon Revolution and Prehistoric Europe* (London: Jonathan Cape, 1973), 292 et seq.

15. Hans E. Suess, "Bristlecone-Pine Calibration of the Radiocarbon Time-Scale 5200 B.C. to the Present," in Ingrid U. Olsson, ed., *Radiocarbon Variations and Absolute Chronology*, Proceedings of the 12th Nobel Symposium, (Stockholm: Almqvist & Wiksell, 1970), 303–13.

16. V. Bucha, "Evidence for Changes in the Earth's Magnetic Field Intensity," *Philosophical Transactions Royal Society*, Ser. A, 269 (1970): 47–55.

17. P. P. Tans et al., "Natural Atmospheric ^{14}C Variation and the Suess Effect," *Nature* 280 (1979): 826–28.

6 PLANTING KNOWLEDGE

1. Charles Dickens, *American Notes for General Circulation* (London: Chapman & Hall, 1842), available at http://facpub.stjohns.edu/~lazrusp/dickens%20reflections%20on%20ny.pdf.

2. Parthenocarpy concerns the reproduction of fruit not preceded by fertilization of the ovules contained in the ovary; parthenocarpic fruits are seedless as are the most commonly consumed bananas today.

3. The genus *Musa* was given its name by Linnaeus in 1753, in honor of Antonius Musa, a Roman physician and botanist, who lived during the reign of Augustus and who remained famous for having saved the emperor's life from a mysterious malady.

4. Luigi Colla, "Memoria sul genere *Musa* e monografia del medesimo/dell'avvocato collegiato Luigi Colla," *Memorie dell'Accademia reale delle scienze di Torino* 25 (1820).

5. Tim Denham et al., "Origins of Agriculture at Kuk Swamp in the Highlands of New Guinea," *Science* 301, no. 5630 (2003): 189–93.

6. Tyler Anbinder, *Five Points: The 19th-Century New York City Neighborhood That Invented Tap Dance, Stole Elections, and Became the World's Most Notorious Slum* (New York: The Free Press, 2001).

7. The Ig Nobels, whose name is a pun on the Nobel Prize and ignoble, are awarded annually for "strange, entertaining, and even absurd research," that "makes people laugh and then think." The purpose of the award is to "celebrate the unusual, honor the imaginative—and spur people's interest in science, medicine, and technology."

8. Kiyoshi Mabuchi et al., "Frictional Coefficient Under Banana Skin," *Tribology Online* 7, no. 3 (2012): 147–51.

9. Supian Bin Samat et al., "The 40K Activity of One Gram of Potassium," *Physics in Medicine and Biology* 42, no. 2 (1997): 407.

10. Simone Moser et al., "Blue Luminescence of Ripening Bananas," *Angewandte Chemie International Edition* 47, no. 46 (2008): 8954–957.

11. *Sgt. Pepper's Lonely Hearts Club Band* is ranked by *Rolling Stone*'s 500 Best Albums of All Time as the "best album in the history of rock." In general, the first five positions are occupied by albums produced between 1965 and 1967.

12. Later on, the United Fruit Company changed its name to Chiquita Brands International.

13. John McMillian, *Smoking Typewriters: The Sixties Underground Press and the Rise of Alternative Media in America* (New York: Oxford University Press, 2011).

14. Abraham D. Krikorian, "The Psychedelic Properties of Banana Peel: An Appraisal," *Economic Botany* 22, no. 4 (1968): 385–89.

15. Stokely Carmichael and Charles V. Hamilton, *Black Power: The Politics of Liberation* (New York: Random House, 1967).

7   PLANTING LAW AND ORDER

1. Gregory Ahlgren and Stephen Monnier, *Crime of the Century: The Lindbergh Kidnapping Hoax* (Tucson, AZ: Branden Books, 1993).

2. Arthur Koehler, "Techniques Used in Tracing the Lindbergh Kidnapping Ladder," *Journal of Criminal Law and Criminology* 27 (1936–38): 712.

3. F. Pope, *State of New Jersey vs. Bruno Richard Hauptmann*, 1935, trial transcript, 3796.

4. T. W. Trenchard, *State of New Jersey vs. Bruno Richard Hauptmann*, 1935, trial transcript, 3805.

5. Grover Maurice Godwin, ed., *Criminal Psychology and Forensic Technology: A Collaborative Approach to Effective Profiling* (Boca Raton, LA: CRC Press, 2000).

6. R. Szibor et al., "Pollen Analysis Reveals Murder Season," *Nature* 395 (1998): 449–50.

7. James Pokines, "Two Cases of Dendrochronology Used to Corroborate a Forensic Postmortem Interval," *Journal of Forensic Identification* 68 (2018): 457–65.

8. FBI, *Handbook of Forensic Services*, 2019, available at https://www.fbi
.gov/file-repository/handbook-of-forensic-services-pdf.pdf/view.

9. Vaughn M. Bryant and Dallas C. Mildenhall, "Forensic Palynology
in the United States of America," *Palynology* 14 (1990): 193–208.

STEFANO MANCUSO is one of the world's leading authorities in the field of plant neurobiology, which explores signaling and communication at all levels of biological organization. He is a professor at the University of Florence and has published more than 250 scientific papers in international journals. His previous books include *The Nation of Plants* (Other Press, 2021), *The Incredible Journey of Plants* (Other Press, 2020), *The Revolutionary Genius of Plants: A New Understanding of Plant Intelligence and Behavior*, and *Brilliant Green: The Surprising History and Science of Plant Intelligence*.

GREGORY CONTI has translated numerous works of fiction, nonfiction, and poetry from Italian including works by Emilio Lussu, Rosetta Loy, Elisa Biagini, and Paolo Rumiz. He is the translator of Stefano Mancuso's two previous books, *The Nation of Plants* and *The Incredible Journey of Plants*, and recently translated *The Child Is the Teacher* by Cristina De Stefano. He is a regular contributor to the literary quarterly *Raritan*.